Environmental Markets

Economics for a Sustainable Earth Series

Environmental Markets

Equity and Efficiency

Edited by

Graciela Chichilnisky
and Geoffrey Heal

Columbia University Press • New York 

Economics for a Sustainable Earth Series
Graciela Chichilnisky and Geoffrey Heal, Editors

Previous publication in this series:
Valuing the Future: Economic Theory and Sustainability, Geoffrey Heal, 1998.

Economic forces are driving dramatic changes in the environment of our planet. Our grandchildren may live in a world radically different from our own. Between our lifetimes and theirs, economic activity may cause changes in climate and in plant, animal, and insect populations greater than any since the evolution of human societies, with far-reaching consequences for human well-being. This poses a challenge to economic analysis, for economists have traditionally taken the economy's material and biological surroundings as given, independent of economic activity. Books in this series grapple with the consequences of human domination of global ecological and biogeochemical systems.

Columbia University Press
Publishers Since 1893
New York Chichester, West Sussex

Library of Congress Cataloging-in-Publication Data

Environmental markets : equity and efficiency / edited by Graciela Chichilnisky and Geoffrey Heal.
p. cm. — (Economics for a sustainable earth series)
Includes bibliographical references and index.
ISBN 0-231-11588-1 (cloth).
1. Emissions trading. 2. Carbon taxes. 3. Sustainable development. 4. Environmental policy. 5. International economic relations. I. Chichilnisky, Graciela. II. Heal, G. M. III. Series.
HC79.P55E584 2000
363.738'7—dc21 99-40305

∞
Casebound editions of Columbia University Press books are printed on permanent and durable acid-free paper.

Printed in the United States of America
c 10 9 8 7 6 5 4 3 2 1

Contents

Preface

This is the second book in a series studying the economic implications of human domination of the planet. The first, *Valuing the Future: Economic Theory and Sustainability* (Geoffrey Heal, 1998), addressed the conceptual issues raised by concerns about sustainability.

Global environmental markets became a timely topic when the 1997 Kyoto Protocol provided a foundation for the development of the first global carbon markets. The protocol is a positive event particularly for the editors of this book, who for many years worked closely with the United Nations Climate Convention and recommended these markets as the institution of choice for the reduction of global carbon emissions.

Several of the articles in *Environmental Markets* are an outgrowth of global negotiations and a product of the lively debate involving the role of distribution and of efficiency in these carbon markets. Because distribution and efficiency are at the core of the relationship between industrial and developing countries, they are key concerns in resolving the thorny issues that have stalled the negotiations for several years. An international meeting organized by the OECD in Paris in June 1993, in which the major players in the global negotiations participated, explored the connections between equity and efficiency. At that meeting, the first-named editor presented findings that are included in this book, namely, that global carbon markets trade unusual goods—global public goods that are privately produced—and that this circumstance leads to an intrinsic connection between distribution and efficiency that does not exist in standard markets. In environmental markets, therefore, the traditional separation of equity and market efficiency may break down. This fact leads to new policy implications for global carbon markets, possibly providing a foundation for win-win solutions in the global negotiations that can benefit both rich and poor countries. Such results may draw together two aims that were previously seen as separate and almost opposed to each other: market efficiency on the one hand and equitable use of the Earth's atmosphere on the other.

As human societies grapple with the increasing scarcity of environmental resources, environmental markets may assume an increasingly important role in the global economy. Along with markets for knowledge, which also involve privately produced public goods, environmental markets are likely to be among the most significant new institutions in the world economy during the next century. Indeed, the new economics of markets with privately produced public goods may provide a means of finding solutions to the global environmental problems.

That some form of equity may be needed for efficiency is a ray of hope in what is otherwise a worrisome environmental picture. Our aim here is to present in a concise and readily available form a compilation of excellent original contributions by a number of experts, which should advance our understanding of how markets contribute to the solution of environmental problems. The articles in this book offer a technical as well as policy-oriented guide for the understanding of the environmental markets of the future.

We wish to thank the authors of the articles as well as many of our colleagues and friends at the various institutions that supported our research during the process of preparing and editing this book. We are especially appreciative of the support from Columbia University, Stanford University, OECD, UNESCO, UNIDO, UNDP and the UN Climate Convention. Greg Howard and Kathy Richardson provided administrative support. Natasha Chichilnisky Heal was a valued supporter during the editors' joint enterprise. Chichilnisky acknowledges the support of Peter Eisenberger, vice provost of Columbia University and director of its Earth Institute and the Lamont Doherty Earth Observatory. To all our friends and colleagues, our warm thanks. To the reader: We hope the work that we started here will grow. It is an important topic and it has a long way to go. We view this book as the start of a journey into the future.

Graciela Chichilnisky
Geoffrey Heal

Chapter 1
Introduction

Graciela Chichilnisky
Geoffrey Heal

Markets are among the oldest and most powerful of social institutions. They are a dominant force in the world economy today and in many ways a force for change and progress. Market economies have led the race for industrialization, overcoming planned economies and traditional agricultural societies during the course of the twentieth century. The most attractive feature of markets is the efficiency with which they allocate resources, requiring minimal intervention once an appropriate legal infrastructure is in place. This was Adam Smith's vision of the "invisible hand" and was formalized in the neoclassical theory of competitive markets that has prevailed in the Anglo-Saxon world since the 1950s.

Since World War II international markets have been remarkably successful. In this period they achieved, to a great extent, a life of their own. World trade increased at least three times more than world production. Even the United States, traditionally an isolated economy, has more than doubled the proportion of trade to economic activity so that international trade today accounts for 30% of gross national product (GNP). The process of industrialization became an irresistible trend in the twentieth century, made global by the dynamics of international markets.

While propelling industrial society forward, markets have also led to excessive use of natural resources. Industrialization to date has been based on energy. It has been, and continues to be, based on the burning of fossil fuels and the attendant emission of carbon dioxide. Scientists now believe that carbon emissions can cause climate change. Economic activity is the fundamental driving force of climate change, and the success of international markets has magnified the use of fossil fuels and other natural resources worldwide.

The international market mediates the relationship between industrial and developing countries, the North and the South. Indeed the developing South specializes in resources that account even today for 70% of Latin American exports and almost entirely for those of Africa, whereas the industrial North specializes in products intensive in capital and knowledge. With few exceptions economic development can be read from the composition of a country's exports. The most successful industrializing nations, the Asian Tigers, have swiftly moved into technology-intensive products and have shaped their markets to fit their development needs.

Since the end of colonialism, international markets have perpetuated a pattern of economic development in which the world's less advanced countries play, to a great extent, the role of resource producers and exporters. This pattern of trade is to some degree explained by the historical difference in property rights between the industrial nations of the North and the developing nations of the South.[1] Countries in the latter hold most resources as common property; in industrial economies these are, on the whole, private property. Differences in property rights have been invoked successfully as a possible explanation of the fact that the South overextracts natural resources for the international market, selling these below real costs (Chichilnisky, 1994). As a result, the North overconsumes resources, and the South overextracts them. In a world where agricultural societies trade with industrial societies, international markets magnify the extraction of resources, and as a result exports of natural resources and their consumption in industrial countries exceed what is optimal.

Almost paradoxically it seems possible that the market institution could solve some of the problems that it helped create. This possibility and the requirements for achieving it are the main themes of this book. The chapters here study the role of environmental markets in moderating today's use of environmental resources. How can markets achieve this goal? The idea is to create and allocate new property rights on the use of environmental resources—local and global—and to allow these to be traded in organized markets. This is an idea in the tradition of Coase, and one of the earliest developments is in Dales (1968). In this book we refer to such markets as *environmental markets*. Environmental markets can operate in many ways. One can trade rights to the use of water bodies or to the use of the atmosphere of the planet for disposing of greenhouse gases. In environmental markets the traders can be individuals or

[1] See Chichilnisky (1994).

corporations. They can also be countries. Such markets already exist in the United States, as permits to emit sulfur dioxide are traded at the Chicago Board of Trade. Following the Clean Air Act of 1990, electric utilities were assigned rights to emit sulfur dioxide up to a specified overall level and were also given the ability to write contracts to trade these rights in open markets.

Because emission markets assign a price to the right to emit, they add a cost to the use of the atmosphere. The cost involved arises either from the need to purchase permits when one exceeds one's allotment or from the opportunity cost of using one's permits allotment rather than selling them at the market price. In all cases environmental markets make environmental resources more expensive and thus discourage their use. Thus, they can induce more rational use of resources globally. This is how markets can help control the overuse of natural resources. Although the idea of using markets to increase the cost of resources is simple, environmental markets themselves are somewhat complex and as yet little understood. The purpose of this book is to advance our understanding of environmental markets so that they can achieve their full potential as a tool of environmental policy.

Two main characteristics separate environmental markets from traditional markets. The first is that environmental markets trade *public goods,* by which we mean goods that are not rival in consumption. An example is the fraction of carbon dioxide in the planet's atmosphere, an amount that is the same for all. The second distinguishing characteristic is that the public goods that are traded are not standard but are privately produced public goods. This means that they are produced by individuals in the course of their everyday lives: By driving cars and choosing to heat our homes, we "produce" atmospheric quality. Thus, environmental markets trade *privately produced public goods.* As the chapters in this book demonstrate, markets with privately produced private goods behave quite differently from standard markets and require a somewhat distinct institutional framework, which is discussed in the following pages.

As a brief background it is useful to point out that the study of markets with public goods goes back to the work of Lindahl, Bowen, and Samuelson (see Atkinson and Stiglitz, 1980), formalized later by Foley (1970) in a general equilibrium context. It is well known that markets with public goods are less efficient than standard markets. Typically, they induce inefficient outcomes. In the case of markets for emission permits, each trader has an impact on everyone's welfare through their emissions, yet their private actions do not take into account the benefits that their emission abatement could produce for others. This miscalculation leads the economy to underinvest in the public good. This might well represent today's problems of global atmospheric quality. Each country benefits the entire world when abating their carbon emissions, yet the benefits they receive are only a fraction of the total, thus leading to less abatement than would be optimal for the world.[2] Markets with public goods lead generally to a less-than-efficient allocation of resources.

To solve this dilemma Lindahl suggested using a different type of market, one with "personalized" prices. In his scheme different traders pay different prices for the public good, depending on their marginal valuations. He showed that when using such prices, markets reach efficient solutions. However, Lindahl's solution is generally considered impractical because one trader can "buy" from another the right to pay less, thus inducing arbitrage among the traders. In the end this can lead to a totally different solution from that intended, one that is no longer efficient. To avoid such outcomes this book remains within the traditional formulation of a competitive market: one good, one price, as opposed to Lindahl-style individualized prices. The chapters in this book study environmental markets that are competitive in the sense that they assign each good one price that is the same for all traders, and no trader has an influence on prices. Although this is a realistic formulation, the problem that Lindahl identified still remains: Efficiency is generally lost when trading public goods in competitive markets. This book proposes a new solution to this dilemma, based on property rights, as discussed below.

A traditional solution that is generally advocated to achieve efficiency in the provision of public goods is for the government itself, rather than market forces, to determine the quantity of the public good produced. However, this solution will not work because the public goods considered here are privately produced. They are produced by individuals in the course of their private lives (e.g., in burning fossil fuels for transportation or for home heating), not by governments. It is not reasonable to expect governments to tell people how much to drive their cars or how and how much to heat and cool their homes, so government determination of the allocation of public goods is impractical in this case, as are personalized prices. Thus, two conventional ways of achieving efficiency in markets with public goods—namely, personalized prices and government choice of the amount of the public good—are not realistic in our case. A new approach is required, and this is a main topic of the chapters in this book.

The chapters in this book look at an alternative way of recovering efficiency in markets with public goods, one that has not been considered until now: the allocation of initial property rights on the privately produced public goods, that is, the rights assigned to the traders before they engage in trading. As a typical example we consider the rights to emit gases into the atmosphere. Re-

---

[2]For more details, see Heal (1994).

cently, such rights have been the subject of policy in the United States (the Clean Air Act allocates rights to emit sulfur dioxide across utilities), and globally (the Kyoto Protocol specifically allocates obligations to reduce carbon dioxide emissions across the industrial nations over a certain period). Rights and obligations are two sides of the same coin and can be used interchangeably in this context. It is widely believed that property rights is an area fraught with social conflict, and to a great extent this is correct. However, in the global environment area these rights are yet to be allocated, so the matter is somewhat open, in contrast with the allocations of, for example, land rights, which are to a great extent already allocated worldwide. Thus, it can be said that it is realistic to consider policies about how rights to use the environment should be allocated. In addition, this is also timely and to a great extent necessary, as the process of allocating rights to environmental use is advancing globally with as yet little understanding of its consequences.

The property rights policies proposed here are especially appealing because they can lead to win-win solutions for all the traders concerned. Indeed chapter 3 in this volume shows that an appropriate allocation of property rights on the use of the atmosphere can lead to efficient allocations in markets in which there is a single price for public goods. This is a somewhat surprising result, as it is well known that single-price markets might not yield efficient solutions in markets with public goods. Furthermore, under certain conditions identified in the articles by Chichilnisky (1993) and Chichilnisky and Heal (1994), the latter reprinted here as chapters 7, reallocating property rights to favor the lower-income countries can make them, as well as the industrialized countries, better off. This leads to so-called win-win strategies and is a result with obvious policy attractions. The discovery of these properties of emissions markets has many intellectual and policy implications, some of which are discussed here and have formed the focus of this book: the issues of equity and efficiency in environmental markets.

As already mentioned the markets considered here are standard competitive markets and as such have a single price for each traded good, or the same price for all traders. The chapters in this book show that competitive markets with privately produced public goods are more complex than standard markets for private goods. Nevertheless, the authors of this book believe that it is worth understanding their properties, because environmental markets are starting to play an important role. They include water markets and markets for trading emission permits. Both air and water quality are privately produced public goods. The destruction of biodiversity by habitat fragmentation and by pollution is also a public good (bad), again privately produced, and the results presented here can apply equally to biodiversity use. As the value of environmental assets becomes more widely understood, markets with privately produced public goods will achieve an increasingly important role.

Other types of markets that also trade privately produced public goods are becoming increasingly important, including markets for the use of intellectual property, such as software products and biotechnology. Knowledge-based goods are similar to environmental assets in that they are privately produced but are nevertheless public goods in the sense that they are not rival in consumption. Thus, markets for privately produced public goods include knowledge markets as well as environmental markets. Both types of markets are likely to play an important role in the decades to come, so it is important to understand their properties and the institutions that are needed to support efficient outcomes. Property rights are important institutions and, as shown here, can be crucial for efficiency.

Markets with privately produced public goods were studied some time ago by Laffont (1977) and others in a partial equilibrium world. This book looks at the problem in a general equilibrium framework, namely, when all markets, for private and public goods, occur simultaneously and interact.

The problems that occupy us here are new, as are the solutions. This book originated from results obtained by Chichilnisky (1993), followed by Chichilnisky and Heal (chapter 7) and Chichilnisky, Heal, and Starrett (chapter 3). These results originated in the context of an OECD policy proposal about global carbon taxes:[3] Chichilnisky (1993) and later Chichilnisky and Heal (1994) addressed the following questions. Given that global emissions of carbon dioxide should be reduced by a certain amount, how should this reduction best be distributed between countries? Should each reduce its emissions by an equal amount? Should the rich countries bear most of the burden? The poor countries? Until these articles were written, it had been a widespread presumption that a given amount of emission abatement would generally have a lower cost in developing than in industrial countries, implying that for efficiency the burden of abatement should be borne disproportionately by developing countries. Underlying this argument was a presumption that the efficient attainment of a given total level of abatement would require the equalization of marginal abatement costs across countries. This would mean that we would start abating where these marginal costs are lower, which was widely assumed to be in developing countries. Thus, in this line of argument developing countries should have been the first to abate and the ones to bear the attendant costs. Chichil-

[3]Chichilnisky acted as an adviser to the Economics Division of the OECD in this context, and Chichilnisky (1993) represents part of the output.

nisky (1993) and later Chichilnisky and Heal (1994) noted that this reasoning is incorrect: Unless unrestricted lump-sum transfers between countries are carried out, Pareto efficiency does not require that marginal abatement costs be equalized. Generally, abatement should take place in the countries that have higher income. Although somewhat surprising at first, the point made by these articles is simple. A dollar to an Indian does not have the same welfare implications as a dollar to an American. So the real opportunity costs of abatement to an Indian might be higher than that to an American even though the dollar cost is lower. Chichilnisky (1993) and Chichilnisky and Heal (1994) went on to show that, even in a world where developing countries can abate at a lower cost, it might still be preferable for industrial countries to abate first.

These results were somewhat counterintuitive to many and led to an interesting debate. Chapter 9 of this book, by Martins and Sturm, addresses this issue. Martins and Sturm seek to clarify the conditions under which one recovers the conventional wisdom that equating marginal costs leads to efficient outcomes and thus that developing nations who have lower abatement costs should abate first. For this they take a different model than the other authors, one in which the consumer's utilities do not depend on the quality of the environment. Within their specific model they show that equating marginal costs leads to efficient outcomes. In particular, if developing nations would have lower marginal costs for abating emissions, abatement should take place first in developing countries. They also show that if in the same model one introduces dependence of utility levels on the public good, Chichilnisky and Heal's results again hold. Thus, the critical issue here is whether the environmental public good affects utility levels directly or only indirectly through its impact on production. In models in which the environmental good has no impact on welfare, the conventional wisdom prevails; in models in which the environment has an impact on welfare, they do not. In general it seems clear that most of the major environmental public goods affect individual utilities; this happens directly, through their health or the amenities available to them, or indirectly, through the climate. Thus, it seems that the conventional wisdom fails precisely in the most realistic models, those in which the environment has an impact on welfare. Indeed, if the environment had no impact on welfare, one might ask somewhat rhetorically, Why bother with environmental policies and with environmental markets?

The background in which these results emerged is as follows. The first results on privately produced public goods in an environmental context addressed the problem of determining which countries should abate carbon dioxide emissions and by how much without, however, containing explicit markets. It was conjectured by some that the lack of markets for environmental goods could be the source of the somewhat unexpected results. Thus, the next natural step was to add competitive markets on emission permits to these models. The same results obtained. The model of an environmental market was formalized first in the Chichilnisky, Heal, and Starrett (CHS) chapter in this book; the rest of the chapters follow this basic model. A main result in the CHS chapter, and indeed the main topic of this book, is a deep connection that emerges between efficiency and the distribution of property rights in markets with privately produced public goods. This is a major departure from standard markets, in which equilibria are always efficient. Here the distribution of property rights matters. It is decisive in ensuring that the market achieves efficient allocations.

Chapter 2, by Chichilnisky and Heal, is a survey of the area of tradable emission markets from the perspective of theory as well as policy. It contrasts the use of carbon taxes with an approach based on trading emission permits and explains the efficiency aspects of markets for emissions quotas. It traces the idea of markets for emission rights to the Coasian view of externalities as arising from an absence of property rights. The first explicit formulation of the idea of a market for emission rights seems to be that of Dales (1968).

Chapter 3, by Chichilnisky, Heal, and Starrett, concentrates on the first welfare theorem in markets in which agents trade, at a uniform price, permits to produce privately produced public goods. The total quantity of permits is taken to be fixed by the government at a level consistent with Pareto efficiency (i.e., at a level equal to that at one of the economy's Pareto-efficient patterns of resource allocation). The article shows that even with the total output of the privately produced public good fixed at a level corresponding to a Pareto-efficient allocation, the equilibria are generally inefficient. This is due to the public good nature of one of the goods traded: the quality of the atmosphere of the planet. What is perhaps more surprising is that, without introducing personalized prices, there exist certain allocations of rights to emit from which the market overcomes the "free rider" problem and achieves efficiency. Thus, equity and efficiency are not divorced as they are in classical welfare economics. This is an important characteristic of competitive markets for privately produced public goods and one that will have significant implications for environmental markets and markets for knowledge.

Chapter 4, by Heal, checks the robustness of the CHS result. It studies second-best optimality in markets with emission permits. Like the Chichilnisky and Heal (CH) chapter, it asks about the optimal pattern of emission abatement across countries, and like the CHS chapter it asks about the performance of emission markets, but now both issues are studied in the context of a total emission level that does not correspond to a Pareto-efficient allocation. In other words it addresses the same issues as CH and CHS, but in a second-best context. It shows that the results for the second-best case are essentially equivalent to the CHS results in the first-best case. To be precise only certain allocations of property rights in emission permits lead to second-best efficiency (i.e., to efficiency subject to the constraint imposed on total emissions). Competitive trading of arbitrary initial allocations of permits does not generally lead to second-best efficiency. Thus, this chapter extends the earlier results of Chichilnisky and CH to economies in which the notion of efficiency is restricted to a second-best environment in which a political process has imposed an emission total not consistent with Pareto efficiency.

In Chapter 5, Heal and Lin delve further into the robustness of the CHS results. They study a different market equilibrium, one in which the traders take into consideration each other's actions and reach a Nash solution, by which each optimizes their choice of abatement given the abatement by all the other traders. Under these circumstances they show that there are generally unique efficient solutions: A unique quantity is abated, and a unique distribution of abatement exists that leads to an efficient solution. In other words the distributional prerequisites for efficiency are even more demanding in the face of Nash behavior.

In chapter 6, Prat discusses an innovative process of allocation of property rights. He postulates a two-stage process in which the traders are given a given share of the total permits first, and then the total amount is chosen. He proves that, with this process, Pareto efficiency can be restored for market equilibrium under certain conditions.

Chapter 7 is a reprint of Chichilnisky and Heal (1994) and is an extension and generalization of Chichilnisky (1993). This article showed that marginal costs will be equal across countries at a Pareto-efficient allocation if and only if the marginal valuations of the private goods are equal in the two countries, a demanding condition that can be expected only with free transfers across the regions. Originally presented in June 1993 at the OECD Conference on the Economics of Climate Change in Paris, Chichilnisky (1993) also establishes that, with Cobb-Douglas utilities, efficiency requires that the fraction of income that each country allocates to carbon emission abatement be proportional to that country's income level. The richer countries should spend proportionally more than poorer nations in abatement. Furthermore, the constant of proportionality should increase with the efficiency of the country's abatement technology. This implies that industrial nations should allocate a larger proportion of their income to abatement. These observations originated a lively interest in the topic of equity and efficiency in environmental markets, parts of which this book documents.

In chapter 8, Hourcade and Gilotte, both of whom were present at the 1993

OECD conference, revisit the original results establishing that efficiency is not connected with the equalization of marginal costs of abatements as in the standard market with private goods.

Martins and Sturm were also present at the 1993 OECD conference, and in chapter 9 they seek to clarify the conditions under which one recovers the conventional wisdom that equating marginal costs leads to efficient outcomes. For this they take a different model than the other authors, one in which the consumer's utilities do not depend on the quality of the environment. Within this model they show that equating marginal costs leads to efficient outcomes. In particular, if developing nations have lower marginal costs for abating emissions, abatement should take place in developing countries. However, they also show that if in the same model one introduces dependence of utility levels on the public good, the CH results again hold: Efficiency might not be associated with equating marginal emission costs. Thus, the critical issue here is whether the environmental public good affects utility levels directly or only indirectly through its impact on production. In general it seems clear that most of the major environmental public goods affect individual utilities directly, through their health or the amenities available to them or through the climate.

Chapter 10 takes the ideas that are central to the earlier chapters, especially that by CHS, and applies them in a different context, namely, the privatization and securitization of the services provided by natural ecosystems. In this case the focus is on a watershed, a case motivated by the decision of New York City to invest several billion dollars in restoring the ecological integrity of its main watershed in the Catskill Mountains. The nontechnical part of this chapter was published as a commentary in the science journal *Nature* (Chichilnisky and Heal 1998). This is augmented here by a formal model of the privatization and securitization process. The relationship with earlier chapters is that many of the services provided by natural ecosystems are privately produced public goods.

Chapter 11 examines further the issue of equity and efficiency in environmental markets. Indicating that in environmental markets a more sophisticated institutional approach is required for efficient market solutions, Chichilnisky proposes the creation of a new global financial institution for this purpose—an International Bank for Environmental Settlements (IBES)—that would combine market features and the voting participation by industrial and developing nations. The proposal was advanced officially at the 1995 Annual Meetings of the World Bank. The IBES mandate would be to obtain market value from environmental assets without destroying them, and it could assist in the organization, intermediation, and regulation of markets for emissions trading, including the borrowing and lending of emissions rights.

In chapter 12, Werksman presents a lucid discussion of the global negotiations that led to the "Kyoto Surprise," the Clean Development Mechanism (CDM) of Article 12. This is the only one of the flexibility mechanisms of the Kyoto Protocol that includes provisions for both industrial and developing nations. Werksman explores the conceptual roots of different aspects of the CDM, including the pilot phase of "activities implemented jointly," and explores the ambiguities in Article 12 and how the CDM could evolve in the future.

In chapter 13, Chichilnisky draws a similarity between markets for knowledge and environmental markets, both of which can be characterized as markets for privately produced public goods. The chapter derives the appropriate models of competitive markets for knowledge and environmental assets and within these markets characterizes the conditions for Pareto efficiency of allocations with privately produced public goods that are expressed in a manner that is independent of the units of measurement and, in certain cases, similar to the Lindahl-Bowen-Samuelson efficiency conditions in the provision of classic public goods.

The agreements summarized in the Kyoto Protocol to the UN Framework Convention on Climate Change (FCCC) are important to the issues addressed in this book. They represent the first agreement to apply market mechanisms to the control of privately produced public goods at the international level in the context of controlling the emissions of greenhouse gases. The driving force behind the success of that agreement was Ambassador Raúl Estrada-Oyuela, who was then chairman of the UN Negotiating Committee of the FCCC. His diplomatic skills were generally agreed to have been critically important in reaching the Kyoto agreement, and in chapter 14 we have a commentary by Estrada-Oyuela on the process leading to the Kyoto Protocol. The text of the protocol follows in the Appendix.

References

1. Atkinson, A. B., and J. E. Stiglitz (1980). *Public Economics.* New York: McGraw-Hill.
2. Chichilnisky, G. (1993). "A Comment on Implementing a Global Abatement Policy: The Role of Transfers." In *OECD: The Economics of Climate Change,* Proceedings of an OECD/IEA Conference, OECD, Paris, June 1993.
3. Chichilnisky, G. (1994). "North-South Trade and the Global Environment." *American Economic Review* 84, no. 4 (September): 427–34.
4. Chichilnisky, G., and G. Heal. (1994). "Who Should Abate Carbon

Emissions? An International Perspective." *Economic Letters* (spring): 443–49. See chapter 7 in this volume.

5. Chichilnisky, G., G. Heal, and D. Starrett. (1993). "International Markets with Emissions Rights of Greenhouse Gases: Equity and Efficiency." Center for Economic Policy Research Publication No. 81, Stanford University, fall 1993. See chapter 3 in this volume.
6. Chichilnisky G., and G. M. Heal. (1998). "Economic Returns from the Biosphere." *Nature* 391 (February): 629–30.
7. Dales, J. H. (1968). *Pollution, Property and Prices.* Toronto: University of Toronto Press.
8. Foley, D. (1970). "Lindahl's Solution and the Core of an Economy with Public Goods." *Econometrica* 38, no. 1: 66–72.
9. Heal, G. M. (1994). "Formation of International Environmental Agreements." In *Trade, Innovation, Environment,* ed. C. Carraro, pp. 301–22. Dordrecht: Kluwer Academic Publishers.
10. Laffont, J. J. (1977). *Effets externes et théorie économique.* Paris: Editions du CNRS.

Chapter 2

Markets for Tradable Carbon Dioxide Emission Quotas: Principles and Practice

Graciela Chichilnisky
Geoffrey Heal

2.1 Introduction

This chapter reviews a range of issues relating to tradable carbon dioxide (CO_2) emission quotas (TEQs). It considers the economic principles on which they are based, compares them with alternative carbon abatement policies, and reviews many aspects of how tradable quotas would be implemented in practice.

Section 2.2 sets the scene, explaining why these issues are on the agenda and how they relate to current issues, such as joint implementation.

The principal alternative to a TEQ regime is the adoption of carbon taxes. Section 2.3 compares salient aspects of the two policy approaches. It also analyzes how they can be combined. Section 2.4 studies a particular and very important aspect of a TEQ regime: the allocation of TEQs among participating countries. These two sections present the key theoretical perspectives on tradable quotas and their main alternative: carbon taxes. Section 2.5 addresses issues connected with the implementation of TEQs, analyzing questions associated with the design and management of a TEQ market.

2.2 The Context of the OECD Discussion

The 1992 Earth Summit in Rio de Janeiro set important goals for the control of the planet's greenhouse gas emissions. Annex 1 countries[1] agreed to roll

[1]Annex 1 countries include the main industrial countries, including the OECD, the former Soviet Union, and the Eastern European members of the former Soviet bloc.

back their emissions to their 1990 levels by the year 2000. Certain institutions share the responsibility for devising policies to implement these goals. These institutions include the Global Environment Facility and primarily the UN Framework Conventions on Climate Change (FCCC), on Biodiversity, and more recently on Desertification.

Industrial and developing countries have rather different perceptions of the issues involved, and these differences are to a certain extent limiting progress in international negotiations. Developing countries fear the imposition of limits to their growth in the form of restrictions on emissions, and therefore on energy use, and more generally on the use of their own resources. They feel that most environmental damage originates currently, and has originated historically, in the industrial countries, whose use of energy and patterns of development are at the root of the environmental dilemmas we face today.

Industrial countries have a different set of concerns. They fear excessive population growth in developing countries and the environmental damage that this could bring about. While recognizing their historical responsibility for excessive environmental use, industrial nations focus on a long-term future in which environmental problems could originate mostly in the developing countries.

In addition to differences in perceptions, scientific understanding of some of the main issues has emerged only recently. Newly found science makes its way slowly into the political decision process because by nature science is highly specialized and is often tentative in its conclusions. The differences in perceptions and the failure to communicate recent scientific findings have hampered the international decision-making process.

2.3 The Economics of the Global Environment

The implementation of the Rio goals of stabilizing emissions at levels not harmful to the climate requires substantial conceptual advances in our understanding of some of the main issues as well as the development of a consensus about the possible policy instruments for tackling these issues. This is not an easy task because the problems of climate change, sustainable development, and protection of biodiversity are all rather new, global in nature, and complex. The economics of climate change involves challenging issues related to economic principles and policies, including the following:

1. The connections among energy use, energy prices, trade, and growth
2. The optimal distribution of quotas to emit greenhouse gases between

countries (As we argue here, the distribution of quotas is not a matter to be judged only on the grounds of equity but can have substantial implications for efficiency.)

3. The conditions that are necessary for carbon taxes to act efficiently
4. The connections among levels of income, optimal property rights, and trading practices in such markets
5. The design of cooperative international policies for the abatement of emissions of greenhouse gases as provided by Clause 4 of the Rio convention

In addition to requiring extensive technical work,[2] implementing the Rio targets requires a deliberate effort on the part of all parties involved to communicate and to understand one another's concerns, to address in depth and critically the problems and the possible solutions, and to reach consensus.

2.3.1 The Present Practice — Joint implementation is a term that is frequently used to describe a cooperative venture between two or more countries to decrease the sum total of their emissions of greenhouse gases. Its origins can be traced to Clause 4 of the Rio convention, which specifically contemplates this possibility. The experience to date has been of relatively small projects involving five countries. One is an agreement involving Norway and Mexico, funded mostly by the Global Environment Facility (GEF). Mexico initiated an effort to replace small electric appliances, such as light bulbs, in a manner that diminishes energy use and carbon emissions. A second project involves the Netherlands in cooperation with Poland and India. Here Poland aims at replacing its use of coal in energy production by natural gas, thereby decreasing its carbon emissions.

In both of these examples, the nature of the cooperation is a bargain between an industrialized country and one or two less developed countries (members of Annex 2 of the Rio protocol), by which the former, in cooperation with the GEF, "purchases" its right to continue its current emission practices through ensuring decreased emissions from the developing countries. The Annex 1 country is credited with an emission reduction that it brought about even though this did not occur on its territory. The experience to date suggests several policy issues that have been the subject of discussion in the FCCC.

[2]These are issues on which recent research at Columbia University and at Stanford University has made much progress (Chichilnisky [3–5], Heal [12–14], Chichilnisky and Heal [2,7], and Chichilnisky, Heal, and Starrett [10]), reflected in the chapters of this book.

2.3.2 The Potential of Joint Implementation — The first, most obvious issue is the effectiveness of joint implementation if taken to its natural conclusion: the purchase by industrialized countries (Annex 1 countries) of rights to continue present emission practices by ensuring decreased emissions from developing countries (Annex 2 countries). Developing countries currently emit at most 30% of the world carbon emissions. Therefore if the aim is to decrease world emissions by, for example, 60% of long-run future emissions, as is often proposed, then even a complete cessation of carbon emissions by all developing countries would at best barely attain this goal. Thus abatement of the type contemplated at present requires active decreases in carbon emissions by industrial countries, which are the main emitters. Joint implementation of the type described here cannot be a substitute.

An argument in favor of joint implementation is that it can lead to improvements in the positions of all the countries engaged in the bargain. This argument is supported by the evidence that the bargain is freely agreed among the countries involved. If countries do not stand to gain, why would they enter the deal? These arguments are correct within a restricted institutional framework but fail to provide a thorough analysis of the situation. What is chosen depends on the alternatives available. A bargain might be better than no bargain at all, but it could be worse than other, alternative bargains that were not within the scope of discussion. With more information about the alternatives available, a country can typically improve its trading position. Indeed the most frequently voiced concern about joint implementation is that a few countries could "steal the march" on others by taking advantage of a thin market with little information. All this is simply a restatement of a well-known fact: Efficient trading requires well-distributed information among all the traders. It also requires competitive trading, which in turn is a function of the number of traders. Two traders typically do not make a competitive market. These two principles, market information and market depth, are widely applied in most well-organized markets across the world and are associated with market efficiency. This leads us naturally to consider a multilateral extension of joint implementation, a framework in which trading is conducted with well-distributed information flows and in which market depth can be achieved through the simultaneous participation of all countries.

2.3.3 A Migration Path to Multilateral Trading? — From the previous remarks emerges another argument in favor of joint implementation. The joint ventures, or "bilateral trading" practices, that characterize joint implementation so far can be viewed as the first step in the development of a well-organized, multilateral market. It is often the case that bilateral trading precedes and leads to multilateral trading. Examples are provided by the Chicago commodity markets and by the Lloyds of London insurance market, both of which started with informal bilateral trading among a few parties. Thus, the challenge is to build a well-defined institutional structure of which joint implementation represents a first developmental step. This requires the construction of a multilateral organization with the clear understanding that today's bilateral joint implementation ventures are to provide data and knowledge about how the multilateral organization will work. The eventual aim is to develop an organization in which countries can achieve an efficient allocation of their resources through decentralized trading by means of well-organized and efficient mechanisms.

2.3.4 Tradable Quotas — A natural multilateral trading organization is a market in which entitlements or quotas to emit greenhouse gases are traded. Such a market has a venerable tradition in economics. At present there are three examples of similar markets in the United States. A sulfur dioxide (SO_2) entitlement market has been trading since the early 1990s on the Chicago Board of Trade. For trading to be possible, property rights must be established. In this case the property rights were established by the Clean Air Act, which restricted the emission rights of the major utilities in the United States. At present trading is conducted mostly between these utilities. Recently, new markets have opened up, as futures and swaps on these quotas have been introduced. These markets are called *derivative* because they trade contracts whose values depend on (are derived from) the value of an underlying asset, in this case quotas to emit. Thus, the prices on these contracts and the gains and losses from trade are derived from the expected prices of the quotas. An electric utility company trades futures because it wants to plan effectively the costs of a projected expansion or reduction of its output, and this will require different quotas from those it holds at present. The next section explains how such markets work to correct externalities and how they can be used to induce a reduction of greenhouse gas emissions domestically and globally.

2.4 Tradable Quotas and Emission Taxes

2.4.1 The Pigou and Coase Traditions — The problem of global climate change addressed by the Rio convention is a classic case of large-scale negative external effects, that is, harmful effects of one party on another that are external to and thus not mediated by the market mechanism. By the emission of CO_2, a country increases the risk faced by all countries, itself included, of a harmful change in climate, thus the existence of a negative external effect. There are two principal approaches to the control and correction of external effects: control and correction through taxes and subsidies, in the tradition established by Pigou [16], and control and correction through the introduction of property rights, as suggested by Coase [3].

Pigou described externalities as stemming from differences between the private and the social costs of an activity. In his vision these differences between private and social costs were to be corrected by taxes or subsidies that alter the private cost of the activity until it equals the social cost. After correction one has the relationship

$$\text{private cost} + \text{tax} = \text{social cost.}$$

Thus, in the case of CO_2 emission, there is a private cost given by the costs of the fuel burned. The social costs include, in addition to the fuel costs, the costs of an increased likelihood of harmful climate change. A Pigovian corrective tax, when added to the private cost, will bring it into line with the social cost.

On the other hand, Coase focused on the fact that goods and services can only be bought and sold and thus brought within the orbit of the market mechanism if they can be owned. Ownership of a good or service means that people can have property rights in these. Coase then saw externalities as arising from an absence of property rights, and, as a consequence, certain economically important goods and services could not be bought or sold and their provision regulated by the market. Thus, in particular the market could not ensure their provision at an efficient level. The natural policy prescription from this perspective is the introduction of property rights for the goods for which they are missing, so that these goods can be traded and their provision regulated by the market. The application of this view to climate change indicates that the services of the atmosphere are being used in the combustion of carbon-based fuels as a depository for CO_2. This happens in a legal framework in which there are no property rights in the atmosphere and thus no opportunity for people to register a demand for it to be left unaltered. In contrast there are property rights over the ground, so that this cannot be used as a depository for waste without permission from the owner, which normally requires payment. Coase's insight is that we need to mimic this situation with respect to access to the atmosphere.

Pigou's insight has given rise to the dominant European policy approach in this field, namely, the use of corrective taxes and subsidies: Coase's has inspired the American approach of tradable permits and quotas, as used in the United States for sulfur dioxide, lead additives, and water discharge rights. The key point in this approach is that before emitting a pollutant into the atmosphere, a firm must own the right to effect such an emission, and such a right is conveyed by the purchase of a TEQ. The creation of these quotas establishes property rights in the atmosphere. If a firm is forced to buy a quota before emitting a pollutant, then, in Pigovian terms, this also raises the private cost of pollution, in this case by the cost of the quota. Once again marginal private costs are changed so that they approach marginal social costs. In fact, in a competitive quota market they will be equated exactly to marginal social costs by the inclusion of the costs of buying quotas.

The two approaches are formally equivalent in important ways, although not in all ways. A tradable quota system requires a polluter to buy a permit before polluting, and this raises the private cost of pollution by an amount equal to the price of the permit. In this respect it appears to the polluter like a tax, as it imposes a tax equal to the price of a permit. Both approaches are consistent with the "polluter pays" principle, which has been adopted by the OECD. Compliance with this is widely viewed as a prerequisite for fairness in the management of pollution. However, from the perspective of the policymaker, there are differences associated with where the main policy uncertainties arise, and we explore these here. There are also differences in the role of the government in each system, as government plays a more central role and of course raises revenue under a tax regime.

2.4.2 Historical Experience and Intellectual Traditions — The different intellectual traditions noted previously lead to different policy regimes, and it is clear that these different intellectual traditions have colored in different ways the policy choices of Europe and the United States.

The Coasian tradition emerged from the University of Chicago, an institution whose influence on economic policy formulation in the United States in the last 20 years has been profound and far-reaching. Thus, the United States has experimented extensively with TEQs in several areas, including the management of SO_2 emissions, management of the distribution of lead additives to vehicle fuels, and management of various emissions in the urban areas of northern and southern California. The United States finds this approach consistent with the prevailing market-oriented approach to economic policy. By the same token tax-based policies have been anathema to a political climate strongly predisposed against taxes, as illustrated by the rapid demise of the Clinton Administration's BTU tax proposal.

In Europe the tradition is quite different. The Pigovian tradition emerged from Cambridge University and is also fully consistent with the French tradition in public economics and economic policy. At the same time most European governments have historically had no natural affinity for market-based approaches to pollution management, having perceived markets as part of the problem rather than as part of the solution. Thus, the concept of a tradable emission quota regime has been less familiar in Europe; rather, the approach that has risen naturally to the top of the agenda is a policy based on carbon taxes.

2.4.3 Uncertainty about Cost-Benefit Relations — One of the main differences between tradable quotas and emission taxes is in the degree of assurance that they offer to the policymaker about the aggregate level of pollution. The point here is simple yet important. It is as follows. With a system of tradable quotas, the aggregate level of pollution is determined to be the total number of quotas issued. If quotas are issued for the emission of, for example, 6 billion tons of carbon dioxide, then, if the system is enforced, the total of emissions will not exceed 6 billion tons. This much the policymaker can be sure of in advance: The total amount of pollution is predictable. However, there is an important aspect of the policy that is not known to her, namely, the cost to polluters of the regulation of emissions to the specified level as measured by the price of an emission quota. This price will be determined by the forces of supply and demand and cannot in general be predicted with any accuracy.

Contrast this with the situation with a pollution tax in which the cost to the polluter is now known with certainty and is of course given by the tax. However, the aggregate amount of pollution cannot be predicted. This will now be determined in the market by the forces of supply and demand. To be precise it will be determined for each firm at the level at which the marginal abatement cost equals the tax on pollution.

With quotas the policymaker is sure in advance of the aggregate amount of pollution that will result from her intervention but is unsure of the resulting costs to industry and commerce. With taxes matters are exactly the opposite: The costs to polluters of policy are known, but the results, in terms of pollution levels, are not. This is a key difference, a key duality,[3] in that in situations of great political sensitivity, knowing the cost of policy intervention to industry and commerce might be essential. This is an argument for taxes. In situations of great sensitivity of the environment to pollution, knowing the aggregate

[3]This duality was first studied by Weitzman [19]. See also Dasgupta and Heal, chapter 13 [11].

level of pollution that will result from a policy might be essential. This is an argument for TEQs.

Threshold Effects

Consider a situation in which the effect of a pollutant on the environment is reversible up to a certain threshold level of pollution that we denote L and is irreversible after that. One can think of many examples. Water bodies can cleanse themselves, provided that they are not "too polluted," but they cannot cleanse themselves if pollution exceeds a certain level. Threatened species can reestablish themselves, provided that their stock is not "too low," but if their stock falls below this level, they are doomed to extinction. Ocean currents and the climates that depend on them remain essentially the same provided that changes in atmospheric temperatures are not "too large," but they can change in a major way if the temperature change exceeds a critical amount.

In each of these cases, there is a level of pollution below which the consequences of pollution are reversible and above which they are not and there is a permanent loss of an environmental asset. It is this threshold level that L denotes. In such situations there is a premium on not exceeding L: The costs of pollution increase sharply beyond L. In such situations there is a strong argument in favor of TEQs, for these can provide the assurance that the aggregate level of pollution will not exceed L. One does this simply by issuing a total of permits that does not exceed L. The only way to reach such assurances with pollution taxes would be to consider the range of all possible marginal emission costs and to pick a tax level that ensures pollution of less than L for any possible marginal emission abatement costs.[4] If the uncertainty about possible marginal abatement cost schedules is great, such a tax might be far greater than is actually needed. In contrast the tax implied by tradable quotas—the price of a quota when the total number of quotas is L—will be exactly the least needed to ensure aggregate pollution less than L.

In many contexts this might be an important consideration in favor of TEQs, as they guarantee that pollution will be within some predetermined limit. There is considerable scientific evidence of threshold effects in the damage that results from many pollutants. All of the previous examples have a real scientific basis.

---

[4]High marginal abatement costs imply high pollution levels for any given pollution tax, as the alternative to paying the tax is cutting back pollution and paying the marginal abatement cost.

Although there are believed to be threshold effects in the relationship between atmospheric CO_2 concentration and climate change, these thresholds are a function of the *stock,* not of the *flow,* of CO_2 into the atmosphere. This means that they depend on cumulative emissions to date and not on the current level of emissions. Cumulative emissions change only rather slowly, and this reduces the importance of the threshold argument in the case of greenhouse gases.

2.4.4 Option Values — The capacity to implement abatement policies in a manner that respects thresholds and so avoids irreversible changes in the physical environment of human societies is an important one in the context of environmental problems in which threshold effects matter. The nature of this importance bears further examination. A key issue here is that we often, indeed usually, do not know how important it is to avoid a change in the environment. For example, we do not know the importance of avoiding major climate changes, nor do we know the importance of preserving certain types of species. Of course we have some ideas, but they are not at all precise, and often they are the subject of disagreement and dispute. Presumably, we will learn more about these as time passes. A quarter of a century hence, our research and experience might have led us to a much better grasp of these issues. In this case it is intuitive that there is a lot to be said for keeping matters as they are until we do know the consequences of a change.

This intuitive point can be formalized in the concept of an "option value" associated with preserving environments as they are.[5] Preserving an environment, say, for 10 years gives us the right and the ability but not of course the obligation to continue preserving it for longer after that. If in 10 years we understand better the consequences of a change, then at that time we can reconsider the preservation issue in the light of better information. Not preserving the environment, irreversibly altering it now, takes away this possibility, the possibility of reviewing our choice in the light of better information. Thus, if we are going to learn more about the importance of environment to society in the future, preserving environments until we have done that learning gives us the possibility of making better-informed long-run preservation decisions. Preservation lets us make a choice when we know more about the possible consequences, and clearly there is a value to this.

The term *option value* is used to refer to this phenomenon because there is the same structure here as is associated with buying an option to purchase

[5]These issues were formalized by Arrow and Fisher [1] and by Henry [15]. This literature is reviewed in Dasgupta and Heal [11] and Chichilnisky and Heal [8].

a security. That option gives you the right, but not the obligation, to buy the security in the future when you have more information about its value. Any policy that maintains the environment, and specifically the climate regime, in its present status quo has to be credited with the corresponding option value. Thus, the existence of the option value is an argument in favor of a conservative environmental policy. In the climate context two conditions are necessary for the option value to be significant: first, that more information about the value of avoiding climate change should become available over time and, second, that climate changes should be irreversible. Both of these conditions appear to be satisfied.

2.4.5 Uncertainty about Future Regulations — A key aspect of CO_2 emission and global climate change is that scientific understanding of this phenomenon is continuously evolving. More is known now than 10 years ago, and the next 10 to 20 years will unquestionably bring even bigger changes. The problems of global climate change might come to be seen as much more or much less threatening than currently. As a consequence of such changes in scientific understanding, the tightness of CO_2 emission regulations will change, becoming more restrictive if the consequences of CO_2 emission are found to be more serious and vice versa.

It follows that there is inevitably uncertainty about the tightness of future regulatory policies with respect to CO_2 emissions. This uncertainty has a cost to firms. For example, when deciding whether to select a technology less intensive in CO_2, a firm will base its decision on the expected costs of CO_2 emission over the life of the project. A utility choosing between oil, gas, and nuclear will make a forecast of the costs of CO_2 emission over the 20- to 30-year life of the project as measured by the costs of tradable CO_2 emission permits or the likely level of CO_2 taxes. In doing so it will recognize the risk of anticipating incorrectly the costs of CO_2 emission and will want to hedge or insure the attendant risk of making the wrong technological choice. An example of such a risk is the risk of selecting a non-carbon-based energy source on the assumption that restrictive emission policies will force up the costs of CO_2 emissions and then finding that in fact a carbon-based energy source is the least expensive and that competitors who have chosen that alternative have lower costs.

An advantage of TEQs relative to carbon taxes is that they can naturally be developed in a way that facilitates hedging this kind of risk. Hedging could occur through the trading of derivatives, such as futures or options on TEQs, a possibility mentioned in previous sections. To elaborate, if a utility anticipates a sharp increase in the costs of CO_2 emission, it will choose the energy source that is least intensive in CO_2 emissions. This exposes it to the risk that scientific research will reveal CO_2 accumulation in the atmosphere to be less threatening than previously believed, with a consequent increase in the number of TEQs issued by regulators and a drop in their price. To offset the risk of being "wrong footed" in this way, the utility would either sell TEQs forward, or buy put options on them. In either event it would profit from a drop in quota prices, and this profit would in some degree offset the costs incurred unnecessarily by the selection of the least CO_2-intensive technology. In the Chicago market for SO_2 emission quotas, utilities have already demonstrated their ability to use such strategies.

2.4.6 Taxes and Quotas: Alternatives or Complements? — Although tradable permits and carbon taxes are generally viewed as the main alternatives in the management of global CO_2 emissions, they are in fact not antithetical. They can be combined in several ways.

Mixed Domestic Policy Regimes

In certain cases a country could find it attractive to employ a mixture of the two approaches. It could have a regime of tradable CO_2 emission quotas but allow firms to emit more than the CO_2 quotas that they hold in exchange for the payment of a tax on each unit of emission in excess of the quotas owned by the firm. For example, if a firm owned quotas to emit 100,000 tons of CO_2 and in fact produces 120,000, then it might be allowed to pay a tax on the 20,000 units by which its emission of CO_2 exceeds the quotas in its possession. In such a regime a firm finding its quota allocation too restrictive would have three options:

1. Reduce emissions
2. Buy more quotas
3. Pay a tax on emission in excess of the quotas possessed

It would choose the least costly option. This clearly implies that the market price of a quota would never exceed the tax rate, for if it did there would be no demand for quotas. One could always achieve the same effect as buying a quota by paying a tax, so that at quota prices above the tax rate there would be no buyers. Thus, the tax rate sets an upper bound on the market price of a TEQ. By setting a tax rate, the regulator bounds the costs to firms of its regulatory policies. This could reduce one of the main disadvantages of a tradable quota regime, namely, the unpredictability of the costs to firms, but at the cost reducing its main advantage, namely, the predictability of the total level of CO_2 emissions. To the extent that a firm can supplement its TEQs by paying taxes, it can in effect create new quotas, making total emissions less predictable.

In a situation in which there is a need for a cap on the cost to industry of a regulatory policy and in which there is also a need for some predictability of the total level of emissions, this mixed system might have a valuable role to play.

Quotas Internationally, Taxes Domestically

Another possible combination of the two approaches is to allocate tradable quotas to countries that can trade them internationally to alter their total allocations of emission quotas and then have countries enforce the given total emission levels domestically either by tax or by command-and-control regimes. In such a system a country that is allocated quotas to allow it to emit 500 million tons of CO_2 might purchase additional emission quotas to bring its total allocation up to 550 million and then implement the national target of 550 million tons domestically by any means it chooses. Of course the commitment to emit no more than 550 million tons would, as already discussed, probably be implemented most accurately by a domestic tradable quota regime, but in principle any domestic policy regime is possible.

2.5 Quotas: Distribution and Efficiency

To introduce a regime of TEQs, we have to create property rights where none previously existed. These property rights must then be allocated to countries participating in the CO_2 abatement program in the form of TEQs. Such quotas have a market value, perhaps a very great one. Thus, the creation and distribution of quotas is potentially a major redistribution of wealth internationally. This of course means that it is economically and politically important, and it is important to understand fully the issues that underlie an evaluation of alternative ways of distributing emission quotas. A clear precedent for this redistributive effect of the introduction of property rights at the international level can be seen in the Law of the Sea conference and the introduction of 200-mile territorial limits in the waters off a nation's coast. The introduction of 200-mile limits established national property rights where none previously existed, and these rights could be and frequently were distributed by governments to domestic firms. The introduction of these property rights in offshore waters effected a very substantial redistribution of wealth internationally.

Clearly, the aim of a TEQ regime is to alter consumption and production patterns internationally. Any policy that is designed to alter global consumption patterns will affect the levels and distribution of consumption. This is especially true in the case of carbon taxes and in the assignment and trading of emission quotas, as both aim at restricting the use of energy, and energy is essential in the production of all goods and services. There is no way to restrict countries' emissions without altering their energy use and so without altering their production and consumption patterns. Thus, the implementation of measures to decrease carbon emissions will have a significant impact on the ability of different groups and countries to produce goods and services for their own consumption and for trade, and the distributional impact of such policy is a matter of major import. This makes their analysis especially difficult because distributional considerations are typically the ones on which consensus is often most difficult to achieve.

The allocation of the world's finite resources among individuals or groups is a central issue in economics, and indeed by itself it practically defines the subject. Market allocations are often recommended on the basis of their efficiency. This means that it is not possible to reallocate resources away from a market-clearing allocation without making someone worse off: There is no slack in the system. Market efficiency requires three key properties of markets:

1. Markets must be competitive.
2. There must be no external effects; that is, in the Pigovian terminology private and social costs must be equal, and in the Coasian there must be property rights in the environment.
3. The goods produced and traded must be private goods, namely, goods whose consumption is rival in the sense that what one person consumes cannot also be consumed by others.

In such markets the outcome is efficient no matter who owns what; that is, the efficiency of a market allocation is independent of the assignment of property rights. Ownership patterns are of great interest for welfare reasons, and different ownership patterns lead to different efficient allocations at which traders achieve different levels of consumption and which are characterized by different distributions of income. However, ownership patterns are of no interest for market efficiency as defined here. The efficiency of the market under these conditions, independently of distribution, is a crucial property that underlies the organizations of most modern societies.

Yet the efficiency properties that make the market so valuable for the allocation of private goods may fail when the goods are public in nature. With such goods it is not possible to separate efficiency from distribution. A good is called *public* when its consumption is not rival, that is, when, to the contrary, what one person consumes is necessarily the same as what all others consume. The atmospheric concentration of CO_2 is a quintessential public good in that it is the same for all of us—we all consume the same amount.[6] Classic examples of public goods are law and order and defense. If these are provided for one member of a community, then they are provided for all.

The public good nature of the atmospheric CO_2 is a physical fact that is derived from the tendency of CO_2 to mix thoroughly and stably. This fact is completely independent of any economic or legal institutions. We can tax emissions or assign rights to emit gases and decide how these can be traded, but nothing changes the physical fact that the atmosphere is a public good. This simple physical fact has profound implications for the efficiency of market allocations. It changes matters to the extent that efficiency and distribution are no longer divorced as they are in economies with private goods. They are in fact closely associated. In economies with public goods, market solutions are efficient only with the appropriate distributions of initial property rights. Why?

It seems useful to argue by analogy, thinking of the market with a public good as far as possible as a market with private goods and checking where the analogy breaks down. This gives us a good idea of the connection between efficiency and distribution in economies with public goods.

A market's operation requires that each trader have a well-defined initial endowments of goods: the traders' *property rights.* This is the same with or without public goods. For example, the property rights in the atmosphere are the trader's assigned rights to use it as a sink for the emission of greenhouse gases. Traders produce and trade goods freely so as to maximize the utility of consumption; the trading activity continues until a market-clearing allocation is reached. Up to this point the analogy between markets with private and those with public goods holds in every sense. However, it breaks down at a crucial point, as market-clearing allocations with public goods can be shown to have very different properties from their private counterparts. This can be seen as follows.

When all goods are private, one expects that different traders will typically end up with different amounts of goods at a market-clearing equilibrium on

[6] Atmospheric CO_2 is an unusual public good in that it is produced privately, unlike centrally produced services, such as defense and law and order. Carbon dioxide is produced by the actions of individuals and firms in choosing the fuels that they use and the amounts that they use. Although we all consume the same atmospheric concentration of CO_2, the implications of this concentration differ from country to country, depending on exposure to the harmful effects of climate change. This does not mean that CO_2 concentrations is not a public good; rather, it means that different countries value this public good differently.

account of their different tastes and endowments. This is indeed the case, and the flexibility of the market in assigning different bundles of goods to different traders is crucial in its ability to reach efficient solutions because, for efficiency, traders with different preferences should nevertheless reach consumption levels at which relative prices between any two goods equal the marginal rate of substitution between those goods *for every trader* and also equal the rate of transformation between the two goods *for every producer.* This is an enormous task to achieve, and it is the decentralized power of markets that must be credited with this coincidence of values at a market-clearing allocation.

However, when one good is public, a physical constraint emerges. All traders, no matter how different, must consume the same quantity of this good, not by choice but by physical laws. It is not possible for traders to consume different atmospheric qualities, even if they want to and even if our economic and legal institutions would allow it. The quality of the planet's atmosphere is one and the same for all traders. This imposes an additional constraint, a restriction that does not exist in markets in which all goods are private. Because of this restriction, some of the adjustments needed to reach an efficient equilibrium are no longer available in markets with public goods.

The number of instruments used by the market to reach an efficient solution, namely, the goods' prices and the quantities consumed by all traders, are the same with private or public goods. However, with a public good these instruments must now perform an additional task. At a market equilibrium the quantities of the public good demanded independently by each trader must be the same no matter how different the traders are. In addition to equalizing prices to every trader's marginal rates of substitution and transformation, one more condition must now be met: The sum of the marginal rates of substitution between the public good and all private goods across all traders must equal the rate of transformation. This condition emerges from the simple observation that one additional unit of the public good produced benefits each and every trader simultaneously. Thus, the physical requirement of equal consumption by all introduces a fundamental difference between efficiency with public goods and efficiency with private goods. All this must be achieved by the market in a decentralized fashion. Traders must still be able to choose freely, maximizing their individual utilities. In other words, with public goods the market must perform one more task.[7]

An additional task calls for additional instruments. Because the market with

[7] Afficionados of economic theory will note that a Lindahl equilibrium provides extra instruments for this task, namely, extra prices, by considering "personalized prices" for public goods. Redistribution of endowments can substitute for the extra prices in a Lindahl equilibrium, as is shown in chapter 3 of this book.

private goods has precisely as many instruments as tasks, with public goods new instruments must be enlisted. Some of the economy's characteristics can now be adjusted to meet the new goals. The traders' property rights on the public good, or their rights to emit gases into the atmosphere, are a natural instrument for this purpose because they are in principle free and undefined until the environmental policy is considered. By treating the allocations of the atmosphere's quotas as an instrument (i.e., by varying the distribution of property rights on the atmosphere), it is generally possible to achieve not only a market-clearing solution but also one by which traders choose freely to consume exactly the same amount of the public good. With public goods market efficiency can be achieved, but only with the appropriate distribution of property rights.

2.5.1 Quota Allocations: North-South Aspects — The physical constraint imposed by the public good is felt most acutely when traders have rather different tastes and endowments. Tastes are often difficult to measure, but differences in endowments are measured readily, as national accounts provide often an adequate approximation. Income differences are very pronounced in the world economy, so that one might expect that the public good problem will have a major effect on market efficiency.

For simplicity one can divide the world into a North and a South, the industrial and the developing countries, respectively. It is fairly obvious that endowments of private goods are much larger in the North than they are in the South; in a competitive market with private goods, this naturally leads to very different patterns of consumption and is likely to emphasize the importance of distributional considerations. Thus, the North-South dimension of CO_2 abatement is likely to be an important aspect in the evaluation of environmental policy. Although this point is widely understood in the context of political negotiations between industrial and developing countries, it has not been clear until recently that the political arguments have in fact an analytical underpinning. Not only are distributional issues fundamental to achieving political goodwill and to building consensus, but, because of the properties of markets with public goods, distributional issues are also fundamental in the design of policies that aim at market efficiency. Market efficiency is crucial in reaching political consensus, as negotiations often advance by producing solutions that are potentially favorable to all. Proposing an inefficient solution means neglecting potential avenues to consensus. This can be a strategic mistake in negotiations in which the achievement of consensus is key.

2.5.2 The Distribution of Quotas among Countries — From the previous arguments it follows that a judicious allocation of quotas among countries must not be viewed solely as a politically expedient measure designed to facilitate consensus. Nor should it be viewed as an attempt to reach fair outcomes at the expense of efficiency or, at least, independently of efficiency. The appropriate allocation of quotas within a given world total of emissions is an instrument for ensuring that competitive markets can reach efficient allocations. The fact that it plays this role derives from the physical constraints that a public good imposes on market functioning. However, what remains to be determined is the particular distribution of quotas that is needed to ensure that the market solution will be efficient. Distributional issues are delicate points in any negotiation and the fact that market efficiency is involved makes the point apparently more complex. However, in reality it can be seen to improve the dynamics of the negotiation process. The reason is that the connection between distribution and efficiency means that an argument about distribution is not a zero sum game, as it would be if all that were involved were the division of a fixed total between competing parties. Because some distributions of quotas are efficient and others are not, some lead to a greater total welfare than others and thus an opportunity for all to gain relative to the other, inefficient distributions.[8] Here we give a conceptual overview of the problem: For applications one needs in additions an analytical framework for computing solutions in each specific case. The latter requires further scientific studies.

Under certain minimal conditions a general recommendation can be reached. We will work under the assumption that all countries have generally similar preferences for private goods and for environmental assets if they have comparable levels of income.[9] This is of course consistent with different trade-offs between private and environmental consumption in countries that are at different levels of income. A second standard assumption is that the marginal utility of consumption decreases with the level of income. This simply means that an additional unit of consumption increases utility less at higher levels of consumption than it does at lower levels: Adding one dollar's worth of consumption to a person with meager resources increases the person's well-being more than adding one dollar's worth to the consumption of a wealthy individual. We assume also that all countries have access to similar technologies and that their productive capacities differ only as a consequence of differences in capital

---

[8]Although we cannot develop this point here, this is true even in a strictly second-best context in which the total emission level being distributed between countries is not one associated with an efficient pattern of resource use overall. In fact, of course, the connection between efficiency and distribution has long been known to be close in the context of second-best policy choices. See chapter 4.

[9]By this we mean only that their income and price elasticities of demand are of the same order of magnitude. We are ruling out radically different valuations of private goods and the environment.

stocks. Under these assumptions an efficient allocation of tradable quotas will require that poor countries be given quotas in excess of their current emission and that rich countries get quotas less than their current emissions.

The previous remarks imply that the allocation of quotas might have to favor developing countries proportionately more than industrial countries if we seek market efficiency. This holds true for any total target level of emissions. However, it seems reasonable to inquire more generally if there is a connection between the distribution of income and the efficient level of emissions reached. To answer this question one must consider one more fact about preferences between private and public goods: that environmental assets are *normal* goods. This is entirely reasonable, as it means that the amount that one is willing to spend on environmental amenities or assets increases with the level of one's income: The more we earn, the more we spend on every normal good, including of course on environmental goods.

The final general condition invoked by our analysis requires perhaps more thought. It is that environmental assets are *necessary* goods. This simply means that whereas the total amount spent on environmental assets increases with the level of income, the *proportion* of income a person is willing to spend on environmental assets increases as the income level drops. This assumption has been corroborated empirically in every known study in the United States, Europe, and Africa, although such studies typically involve contingent valuation techniques, which can have weaknesses.[10] The assumption can be theoretically justified on the grounds that lower-income people are more vulnerable to their environment than are higher-income people. The latter can afford to choose or modify their environment, whereas the former cannot. For example, public parks or access to potable water are environmental assets that have relatively more value to lower-income people than they do to those who can afford to build their own parks or arrange their own water access. Humans in lower-income countries are known to be more vulnerable to the effects of global warming than those in higher-income countries. Thus, we propose a plausible formulation of a fact that has been established with remarkable regularity in all known empirical studies, namely, that the income elasticity of demand for environmental assets is between zero and one.

From these facts it is possible to establish that a redistribution of income toward lower-income individuals or countries will generally lead to an improvement in the world's emission levels and in the world's level of environ-

[10]This has now been documented in a large number of studies in many different countries. A good reference is a paper by Kristrom [16].

mental preservation. This is because when preferences are similar and the income elasticity of demand is less than one, a redistribution of income in favor of lower-income groups implies that relatively more income will be allocated to the environmental asset. If traders choose freely, they will choose more preservation. In our case higher abatement levels are to be expected when more resources are assigned to the lower-income groups of countries.

However, there is another factor that must be considered. Developing countries could be less efficient in terms of energy use and thus lead to more emissions as they grow. This is certainly an important concern for the long-run future, that is, 50 years or so. Indeed it seems that such concerns should drive environmental policy today. Every effort must be made to help prevent developing countries from adopting the patterns of environmental overuse of industrial countries as they grow.

2.6 The Design of the Market

2.6.1. Transaction and Implementation Costs — Any policy has certain implementation costs associated with it. These are rather different in nature for the two policy alternatives under review here. For a tradable quota regime, the costs are as follows:

1. The costs of establishing and maintaining a market
2. The costs of transacting in the market
3. The costs of monitoring and ensuring compliance with the policy

For a carbon tax regime, one has the following as cost categories:

1. The costs of collection
2. The costs of monitoring and ensuring compliance with the policy

Costs of a Tradable Quota System

The costs of establishing and maintaining a market are fixed costs, that is, costs that are largely independent of the size of the market and the volume of business conducted in it. An effective market requires a legal and contractual framework that defines the commodity to be traded, establishes the contractual obligations of the parties to a trade, and sets out payment and settlement mechanisms. The costs of establishing such a framework are likely to be large in the first place. Because they are independent of the volume of transactions, they will be substantial on a per trade basis for low trading volumes but will become quite acceptable per trade if, as seems likely, the volume of transactions eventually rises to several $U.S. billion per year. Thus, they are probably not a major factor in the choice of policy regime, although it must be emphasized that a successful market does require regulation and a good legal infrastructure.[11]

The costs of transacting in the market, of buying and selling, depend on the nature of the market and on its liquidity. In some tradable quota markets, these have been quite high: Stavins [17] cites a figure in some cases as high as $25,000 per transaction (on transactions that are valued at millions of dollars). Such high figures occurred because prior to the development of the Chicago SO_2 quota market, markets were decentralized and operated through brokers acting as intermediaries. The role of the brokers was to bring together buyers and sellers, so that they claimed an introduction fee as well as a buy-sell spread. The transaction costs on the Chicago market are now very much less and are of the same order as transaction costs in organized financial markets. Such costs are low enough not to be a major factor in the evaluation of a tradable quota regime.

There is one important general observation about the costs of TEQ regimes: There is a trade-off between the size of transactions costs in the market and the level of the initial investment in market infrastructure. The point here is that the larger is the initial investment in establishing a transparent well-run market open to all would-be traders, the smaller are the per transaction costs when the market is operating. The reason is that a well-run centralized market obviates the need to pay brokers and other intermediaries to find counterparties to a transaction. It also greatly reduces the costs of settling a transaction and, by providing a standard legal framework and establishing contractual relationships between trading parties and the market, reduces the risks associated with possible failure of a counterparty to a trade to perform their part of the deal. In informal markets characterized by bilateral bargains, these risks have historically been considerable. A well-run market provides a supply of traders, a contractual framework that minimizes nonpayment and nondelivery risks, and an organized payment-and-settlement system.

The costs of monitoring and verifying compliance are much the same under either policy regime. These are the costs of verifying that a quota is opened or a tax paid for each unit of CO_2 emitted. As discussed in the following, this will typically not require the measurement and recording of each unit of CO_2

[11]The securities markets of the United States, by general agreement the most active and open in the world, are heavily regulated and managed by the Securities and Exchanges Commission.

emitted. A compliance system will typically require quotas to be purchased or taxes paid at the wholesale level. It will require producers of gas, coal, and refined petroleum products to comply with the provisions of a tradable quota or tax regime on the principle that the overwhelming majority of carbon-based fuels to reach end users will pass through these channels. Such an approach will limit the number of sources to be monitored to a number in the hundreds or, at most, thousands.

The Costs of a Tax Regime

The infrastructure needed to implement a regime of carbon taxes is quite conventional relative to that required for a tradable quota regime and is of a type already possessed by almost every government. It is essentially the administrative apparatus need to administer a fuel tax, which is already in place in many countries. The costs of monitoring compliance with a tax regime are the same as those of monitoring compliance with a tradable quota regime and have already been discussed.

Private Sector Involvement in Implementation

The governments of most countries will probably find it easier to implement carbon taxes than tradable quotas. However, it is possible that private-sector financial institutions will be willing to organize and provide much of the institutional framework needed for a tradable quota regime in exchange for the right to participate as brokers and market makers in the resulting markets. In financial markets such rights to participate are valuable, as in many cases the markets are financed by charging membership fees to the financial institutions who subsequently become the key participants. Several major private-sector international financial institutions have already indicated interest in becoming participants in a global CO_2 TEQ market.

2.6.2 The Organization of Quota Markets — For the full economic potential of a regime of TEQs to be realized, the market for tradable quotas must be competitive and free of manipulation and should give all would-be traders equal access to information. It must also provide mechanisms for hedging price uncertainty. The issue of hedging mechanisms is addressed in section 2.5.6 and in Chichilnisky [6].This section focuses on issues associated with the nature of competition in quota markets and the organization of access to these markets.

A key issue is whether the number of traders in these markets will be large enough to ensure competition and whether any of the traders will have the power to dominate the market. These issues are in turn linked to the question of who participates in the tradable quota markets. There are several possibilities here, and mixtures of them are possible as well:

1. International quota markets will be intergovernment markets, purely for the redistribution of quotas between countries. Participation would be open only to designated government agencies.
2. International quota markets will be open to all firms in all countries, establishing a truly global market for CO_2 emission quotas.
3. The international market will be open only to governments but will be supplemented by domestic quota markets within which firms in a country trade the quotas that have been issued to or purchased by its government.

In terms of establishing a truly competitive market, the second option here—international markets open to all comers—would be the best. However, such an approach would raise questions about the ability of governments to implement national policies, as it would allow the transfer of permits between countries without any government approval.

The issue of whether firms in a country should be freely able to export or import tradable quotas is a complex one. Many governments will have an instinctive reaction to restrict this ability and retain control of the total number of quotas in their country. There would in fact be no reason for restrictions on the export or import of quotas if and only if it were clear that market prices reflected fully the social value of a tradable CO_2 quota to a country. In this case the export of an emission quota from a country would give it an amount of cash that fully compensated for the loss of the quota.

Unfortunately, there are likely to be many circumstances in which this condition is not fulfilled. For example, a developing country's government might feel that the current market price of an emission quota does not reflect the value to it of that quota at some future date when its industrialization strategy is further advanced and its emissions of CO_2 consequently much greater, and thus it might want to accumulate quotas not currently needed for future use. An alternative strategy, feasible if there is a liquid futures market for quotas, would be for the country in such a position to allow the sale of current quotas and at the same time to make forward purchases to cover anticipated future needs.

In an active market one would expect to see "maturity swaps" developed to provide precisely this service. Equivalent swaps are routine in government debt markets and are also available in the Chicago market for SO_2 quotas, in which a utility with a surplus of quotas for the near future and a deficit for the longer term may swap the surplus near-term permits for permits of future validity.

There are several possible models of what might ultimately emerge if a global tradable quota policy is adopted. One is a two-tier market system. In this case one might see regional markets in such areas as North America, western Europe, and South America, with all firms and governments in a region free to trade on the regional market, and then a global market in which only governments or regional authorities trade to alter the distribution of quotas between regions.

An alternative would be a global market in which some governments allow domestic firms to trade directly on the global markets and export or import quotas freely and in which other governments restrict the right to trade on the global market. In such a case the major industrial countries might be expected to permit any domestic firms to trade on the global quota market, whereas developing countries' governments might exercise more control over the import and export of quotas. For example, they could impose tariffs on trade in quotas, requiring exporting firms to pay a fraction of the revenues from exports into a national tradable quota bank, or require export licenses.

From the perspective of ensuring a competitive market with incentives for brokers to innovate in the production of instruments such as swaps, futures, and options, the last regime is clearly the best.

2.6.3 Design of the Tradable Quota — What exactly is the object to be traded in a market for tradable emissions permits? The fundamental source of possible climate change is the stock of CO_2 in the earth's atmosphere. The larger this is, the larger is the chance of a significant change in the climate. Thus, the ultimate objective of economic policies is first to stabilize and then to reduce the stock of CO_2 in the atmosphere. There is a natural CO_2 cycle in the environment by which human activity emits CO_2, which is removed from the atmosphere either by solution in the oceans or by photosynthesis by green plants or by microorganisms in the ocean. This process turns CO_2 into energy for plants and microorganisms and into oxygen, which is emitted into the air. Thus, to stabilize and then reduce the stock of CO_2 in the atmosphere, the emission of CO_2 has to be reduced below the rate at which it is removed from the atmosphere by solution in the oceans and photosynthesis. One part of a policy strategy might be to increase the rate of removal by photosynthesis, which can be affected by the preservation and extension of forests. In principle, then, a policy has to discourage the emission of CO_2 and encourage its absorption.

What are the implications of this for the nature of tradable quotas? Damage inflicted depends on the stock of CO_2 in the atmosphere and not on the flow of CO_2 into the atmosphere. The rate of emission of a given total is much less important than the size of the total. It is of limited concern whether a given amount of CO_2 is emitted at a great rate over one month or much more slowly over a year or more. Thus, quotas should govern the total amount of CO_2 to be emitted over some interval, not the rate of emission. This means that a five-year quota for, say, 100,000 tons of CO_2 entitles the holder to emit a total of 100,000 tons in any time pattern whatsoever over the five-year validity of the quota. It is not a right to emit 20,000 tons annually for five years. The 100,000 could all be emitted in the first month, or in the last month. The timing of emission might matter only in one respect, namely, that the social costs imposed on the global community by an incremental unit of emission might be less when the stock of CO_2 in the atmosphere is less. In the limit, if the stock in the atmosphere were to return to preindustrial levels, there would be no social costs of emission not reflected in the private costs. However, within the foreseeable future this is likely to be an insignificant effect, and it seems safe to assume that within 5- to 10-year intervals the timing of emission is irrelevant to the economic significance of the emission.

However, from the perspective of a firm, there are important issues related to the timing of the emissions allowed by a quota and the duration of the quota. A firm seeks to choose the least-cost technology for a certain purpose. Suppose, for example, that a utility selects oil as the least-cost energy source on the basis of present and anticipated energy prices and prices of CO_2 emissions permits. Then, by constructing an oil-fired power station, it will be making a 20- to 30-year commitment on the basis of these prices and will want to "lock in" these prices to the greatest degree possible. In the case of emission quotas, this could be facilitated by the regulatory regime in one of two ways. One way is to give quotas a 20- to 30-year life, so that quotas purchased now by the utility at current prices will remain valid over the life of the power stations that it intends to build. An alternative way is to give shorter life spans to the quotas, perhaps 5 to 10 years, but establish futures markets in quotas so that the utility can lock in a supply of quotas for the life of its power station today at known prices.

From the regulatory perspective, there is a difference between these two approaches, that is, between giving long-lived quotas or establishing futures markets in shorter-lived quotas. The latter approach gives more flexibility. In particular it allows changes in the distribution of quotas. As discussed in chapters 3, 6, 7, 8, 11, and 13 of this volume, the allocation of CO_2 emission quotas between countries is a politically complex and important issue, and it is quite possible that it might be appropriate to alter this allocation over time, for example, by shifting the distribution of quotas over time toward the developing countries. If quotas have a life of 20 years, a distribution cannot be changed within this time span. If they have a 10-year life, then after 10 years a new set of quotas can be distributed according to different rules. One remark that should be made about this possibility is that if there are short-lived quotas and uncertainty about the future distribution of quotas, this would lead to uncertainty about the future prices of quotas. Countries uncertain of their future allocations would not know whether they would be net buyers or sellers, so that future prices could not be established. Thus, if quota distributions were to be altered over time, it should ideally be according to a preannounced strategy.

2.6.4 Enforcement Framework

Monitoring Compliance

There are two aspects to an enforcement framework. One is the monitoring of compliance with the regulatory framework and detecting violations. The other is responding to violations in a way that ensures that it is always in the interests of participants to comply.

The first of these aspects is by far the more straightforward of the two. Arrangements for monitoring compliance have been mentioned several times. In particular, we have made the point that to monitor overall compliance it is not necessary to monitor every possible industrial source of CO_2. It will be sufficient to monitor the sales of the major distributors of carbon-based fuels (i.e., the major distributors of gas, oil, and coal). These are limited in number and fairly prominent. Provided that the sales of fossil fuels by these agents are within a country's quota, the total use of such fuels must also be within the quota. These distributors are of course not the ultimate users of fossil fuels and so are not responsible for burning them and emitting CO_2. Thus, they would not be required to hold permits, but nevertheless their outputs would provide a good guide to the total emissions of CO_2. The TEQs would be needed and traded by their customers. In fact, estimates of the consumption of the various carbon-based fuels in each country are already available from data on production, import, export, and inventories. Such data are available to international agencies and would be difficult to falsify to a significant degree.

It is also possible to monitor fairly readily the preservation of carbon dioxide sinks, such as forests and other large areas of vegetation. The extent of these can be observed and measured from satellite pictures. In fact these are the main sources of internationally agreed data in this area today.

Thus, there is the capacity to monitor annual emissions and absorptions of CO_2 by countries. However, as noted in the previous section, emission quotas should not in general specify an annual emission rate; rather, they should specify a total of emissions over a multiyear life. If all the quotas in a country have the same validity dates—for example, all are valid from 1995 to 2005—then this does not complicate matters, as it is decadal rather than annual emissions that are monitored.

If the lives of quotas are not synchronized, matters could be more difficult. Consider, for example, a country with two utilities using quotas. One has a quota valid from 1995 to 2000 for a total emission of 0.5 million tons and a quota valid from 2000 to 2010 for 1 million tons. The other utility has a 1-million-ton quota from 1995 to 2005 and a 0.5-million-ton quota from 2005 to 2010. In this case emission from 1995 to 2000 could legitimately be anywhere in the range from 0.5 million tons to 1.5 million tons. The upper end of the range would occur if the second utility used all of its quota for 1995 to 2005 in the first five years of its life. It is probable that with large numbers of quota-using firms, such effects would be less significant in the aggregate. It is also likely to be the case that the lives of quotas will be synchronized.

Enforcing Compliance

The enforcement of an international agreement clearly poses serious problems, although there are many precedents for multinational agreements that have been respected by their signatories. These include the Montreal Protocol on Substances that Deplete the Ozone Layer and the Nuclear Non-Proliferation Treaty, both of which limit either environmental emissions or national sovereignty over power sources and thus have some element in common with a treaty on global warming. However, a global warming treaty would be much more far-reaching than either of these.

Ultimately, enforcement could be achieved only by a combination of enlightened self-interest and diplomatic and economic pressures, as the international community has no effective legal sanctions that could be used to ensure compliance. Economic pressures would be exerted through international agencies and patterns of international trade and diplomatic pressures through the usual diplomatic sources. The successful implementation of a broad-based global warming treaty would unquestionably pose new challenges to the international community and set an important precedent for planetary cooperation on environmental matters. Successful implementation is related to the nature of the countries that agree to participate in the treaty. In the next section we argue that the incentives to comply increase with the number of participants, and indeed that with sufficient participation compliance, will be in each country's self interest.

2.6.5 Participating Countries — How many countries, and which countries, have to ratify a global warming treaty for it to be worth implementing in the sense that it will make a real difference to the threat of climate change? Perhaps more important, how many countries have to ratify such a treaty for all the signatories to feel that they will gain from the treaty and that it justifies their support and commitment? This is closely related to the issue of enforcement discussed in the previous section.

There are several analytical issues behind these questions. A global warming treaty is unlikely to have the participation of all countries as soon as it starts; rather, it is likely to begin with limited participation and to gain support over time. Thus, the group of countries that starts the treaty must be such that they all feel that the group is durable and that the group will continue to abide by the treaty for long enough for widespread support to build up. Whether this condition is met depends very much on the size and composition of the initial group.

A key issue here is that the gains to all countries from participation in a global warming treaty depend on and increase with the number and size of the participating countries. The costs to each country of participation also fall as the number of participants increases. There is a sense in which there are economies of scale in the formation of such agreements. There are two key points here.

One is that when a country cuts back its emission of CO_2, it alone pays the costs of this abatement; however, benefits accrue to all other countries that would be negatively affected by climate change, because climate change, if it occurs, will be worldwide. It follows that if one country abates CO_2 emission on its own, it will clearly be a net loser from this, as it will meet all the costs, and many other countries will share the benefits with it. Suppose, however, that a group of countries agree jointly to abate carbon emissions. The costs of each country's abatement, as before, are borne by that country, but each country now gains not only from its own abatement but also from that of all the other participating countries. The ratio of benefits to costs is now much more favorable. The costs to each country are unchanged, and the benefits to each country are multiplied by the number of participating countries.

In fact, and this is the second point leading to scale effects in the formation of abatement agreements, countries' costs might actually be reduced if the abatement is part of a simultaneous policy move by several countries. One of the main costs of CO_2 abatement is the development of new technologies, and if this is done collaboratively by several countries, each might face a lower individual abatement cost. There is clear evidence of this in the case of unleaded vehicle fuels. Once refining practices and engine designs to cope with these had been developed in the United States (at considerable costs), these technologies could be deployed by the companies that developed them in other countries at little or no incremental cost.

It follows from this that there is a "critical mass" issue in forming the initial group of signatories of a CO_2 abatement treaty.[12] The group has to be big enough (size here is measured in terms of the fraction of global CO_2 emissions controlled) that the gains to each country from participation of the others are sufficient to outweigh the costs that each country incurs. Once such an abatement configuration is in place, problems of deliberate noncompliance at the national level should be greatly reduced.

Another analytical issue in evaluating the adequacy of a group of signatories to a global warming treaty is the phenomenon of "carbon leakage." This refers to the fact that if there is agreement by a group of countries that are major energy consumers to cut back the use of fossil fuels as part of a CO_2 abatement policy, then the consequent decrease in their demand for these fuels will decrease their prices on world markets and so encourage other nonparticipating countries to consume more. This could partially offset the policies implemented by the signatories of the global warming treaty. There is as yet little agreement about the possible magnitude of the phenomenon of carbon leakage,[13] and indeed there are several other mechanisms through which leakage can occur.

What are the implications of these issues for the group that should be targeted as the initial signatories of a CO_2-abatement agreement? Such a group has to be sufficiently broad based to meet two conditions:

1. It has to form a critical mass in the sense of being large enough to ensure that all members gain from membership and so have incentives to remain in compliance.
2. It has to be large enough that the carbon leakage phenomenon does not detract from its efficacy.

However, it need not contain initially all the countries that will ultimately have to join to make it a complete success. It should certainly contain the major industrial countries—the members of the OECD. The additional groups who will ultimately have to join for complete success are the economies of eastern Europe and the former Soviet Union and the major developing countries, such as India and China. It is probably not necessary for all these additional countries to be full members of a global warming treaty as soon as it starts, as long as two conditions are fulfilled:

[12] This point is developed in Heal [2,3].

[13] A more detailed discussion of these effects can be found in OECD Economic Studies, No. 19.

1. They will not pursue policies that will undo the efforts of the signatories of a global warming treaty; that is, they will not increase their emissions of CO_2 to offset, fully or partially, the measures taken by signatories. In particular, they will neutralize carbon leakage.
2. They express an intent to participate fully within a specified period of, say, 10 years.

In fact these aims could easily be achieved by all countries joining a TEQ regime if the OECD countries were allocated quotas that forced them either to reduce emissions or to buy from other countries and if the developing countries were allocated quotas sufficiently in excess of their current needs that they would not constrain their economic development in the near future. In effect the developing countries would then be sleeping members of the treaty for a period but during this period would be able to benefit from the sale or loan of their excess quotas to industrial countries, providing them with an incentive to keep carbon emissions low and maximize the revenues obtainable from quotas. Such a distribution of quotas is, as already noted, consistent with their efficient allocation.

2.6.6 Market Management

Instruments for the Trading of Emissions Quotas

What instruments, apart from the basic tradable quotas, should be traded on the markets that form a part of a tradable quota regime? The role of derivatives such as futures and options in facilitating hedging price risks has been mentioned several times and clearly is important. These instruments, plus various maturity swaps, are already traded in association with the SO_2 quota market on the Chicago Board of Trade, where experience to date confirms the importance of these instruments in hedging.

There is an additional argument for the introduction of such products, namely, that derivatives help achieve market depth and liquidity and so improve market functioning. In the market they serve two important functions. They reallocate risks, as do all financial instruments, and they function as substitute credit markets, allowing traders with limited liquid assets to trade extensively. For example, trading options on oil futures requires less cash than trading oil futures. Thus, market liquidity is increased with options.

Borrowing and Lending versus Buying and Selling

So far we have spoken entirely in terms of the purchase and sale of emission quotas: sale by countries with a surplus over their immediate requirements and purchase by those whose emissions exceed their allocation of quotas. It is clear that some countries feel an unease at selling, parting permanently with their rights to emit greenhouse gases, rights that they might need in the future at a different stage of economic development. In principle they can of course buy these rights back in the future when they are needed, although there is a risk that the price will then be excessive. This risk can, as already mentioned, be reduced by the use of futures contracts or maturity swaps. Nevertheless, there might remain a residual unease about the sale of emission rights. There is a rationale for this, as no one can predict the liquidity of the TEQ market or the prices in that market several decades hence.

An alternative approach is to allow countries to lend or borrow emission rights rather than buying and selling them or indeed to allow both. We can conceive of a central bank[14] at which quotas are deposited when not needed and from which deficit countries borrow quotas. A country with a surplus of permits that it anticipated continuing for, say, five years would make a five-year deposit in the bank and be paid interest on this deposit. After five years it could withdraw its permits or roll over the deposit. Through this system a country's total emission rights never change: It never gives them up permanently but simply lends them while they are not needed.

The interest rate payable on permits would of course depend on the balance of supply and demand for permit loans. A large number of would-be borrowers with few lenders would force up the interest rate and vice versa. The interest rate would be affected strongly by the initial distribution of permits.

Such a system not only bypasses the reluctance that countries might feel with respect to selling emission quotas but also reduces the risks in the market because each party would be dealing with an international institution—an international environmental bank—which would have a credit status similar to that of the International Monetary Fund (IMF) and the World Bank. This arrangement would remove any counterparty risks linked to trading with countries of uncertain credit worthiness.

References

1. Arrow, K. J., and A. C. Fisher. "Preservation, Uncertainty and Irreversibility." *Quarterly Journal of Economics* 87 (1974): 312–19.
2. Coase, R. H. "The Problem of Social Costs." *Journal of Law and Economics* 3 (1960): 1–44.

---

[14]Elsewhere, Chichilnisky has written on the case for a "Bank for Environmental Settlement" that could play this role. See chapter 11.

3. Chichilnisky, G. "A Comment on Implementing a Global Abatement Policy: The Role of Transfers." Paper presented at the OECD Conference on the Economics of Climate Change, Paris, June 1993. In *OECD: The Economics of Climate Change,* T. Jones, ed. (Paris: OECD, 1994).
4. Chichilnisky, G. "North-South Trade and the Global Environment." *American Economic Review* 84, no. 4 (September 1994): 851–74.
5. Chichilnisky, G. "What Is Sustainable Development?" *Land Economics* 73, no. 4 (November 1997): 467–91.
6. Chichilnisky, G. "Markets with Endogenous Uncertainty Theory and Policy." *Theory and Decision* 41 (1996): 99–131.
7. Chichilnisky, G., and G. M. Heal. "Who Should Abate Carbon Emissions: An International Perspective." *Economics Letters* (Spring 1994): 443–49.
8. Chichilnisky, G., and G. M. Heal. "Global Environmental Risks." *Journal of Economic Literature* (fall 1993): 65–86.
9. Chichilnisky, G., and G. M. Heal. "Markets for Tradable CO_2 Emission Quotas: Principles and Practice," Economics Department Working Paper No. 153, OECD, Paris, 1995. Also published as chapter 10 in *Topics in Environmental Economics,* ed. M. Boman et al. (Amsterdam: Kluwer Academic Publishers, 1999).
10. Chichilnisky, G., G. M. Heal, and D. A. Starrett. "Equity and Efficiency in International Permit Markets." Working paper, Stanford Institute for Theoretical Economics, Stanford University, fall 1993.
11. Dasgupta, P. S., and G. M. Heal. *Economic Theory and Exhaustible Resources.* Cambridge: Cambridge University Press, 1979.
12. Heal, G. M. "International Negotiations on Emission Control." *Economic Dynamics and Structural Change* 3, no. 2 (1993): 223–40.
13. Heal, G. M. "Markets and Biodiversity." Paper presented at a conference on Biological Diversity: Exploring the Complexities, Tucson, Arizona, June 1994.
14. Heal, G. M. "Valuing the Very Long Run: Environment and Discounting." Working paper, Columbia Business School, Columbia University, fall 1993.
15. Henry, C. "Option Values in the Economics of Irreplaceable Assets." *Review of Economic Studies* (1974).
16. Kristrom, B. "Is the Elasticity of Demand for Environmental Goods Less Than One? Evidence from International Studies." Paper presented at the Second Ullvon Conference on the Environment, June 1994.
17. Pigou, A.C. *The Economics of Welfare.* London: Macmillan, 1932.

18. Stavins, R. "Transactions Costs and the Performance of Markets for Pollution Control." Working paper, Kennedy School of Government, Harvard University, fall 1993.
19. Weitzman, M. L. "Prices vs. Quantities," *Review of Economics Studies* 41, no. 4 (1974): 477–91.

Chapter 3

Equity and Efficiency in Environmental Markets: Global Trade in Carbon Dioxide Emissions

Graciela Chichilnisky
Geoffrey Heal
David Starrett

3.1 Equity, Efficiency, and Carbon Dioxide Abatement

This chapter addresses a topical issue: the creation of a global market for carbon dioxide (CO_2) emission permits.[1] The recent adoption in the Kyoto Protocol of an ambitious target for global CO_2 emission has focused attention on policy instruments for achieving this goal. In addition, increasing awareness of the economic burden of environmental protection has produced an interest in market-based policy instruments that can minimize detailed government intervention. As a result markets for emission rights are today the approach of choice of the U.S. administration.[2]

This chapter is based on Chichilnisky, G., Heal, G., and Starrett, D. "International Markets with Emissions Rights of Greenhouse Gases: Equity and Efficiency," Center for Economic Policy Research Publication No. 81, Stanford University, Fall 1993.

[1] The atmospheric concentration of CO_2 has become a matter of international concern. It is generally recognized that it has the capacity to change the global climate in ways that are potentially harmful and irreversible. For a review, see Chichilnisky and Heal [3] and Chichilnisky et al. [8]. Consequently, countries at the 1992 Earth Summit in Rio de Janeiro agreed to cut back CO_2 emissions to their 1990 levels by the end of the twentieth century. This policy could easily cost several percent of GNP (see Weyant [27]). In conformity with the conclusions of the Earth Summit, the U.S. administration has recently made a tentative move in the direction of capping CO_2 emissions in industrial countries.

[2] According to a statement by Tim Wirth, U.S. assistant secretary of state for global affairs, at the 1996 Berlin Conference of the Parties of the Framework Convention on Climate Change.

We show that a market for emission permits has an important characteristic not previously noted, a characteristic that has significant economic and political implications. When the level of emissions affects utilities, there is an unexpected link between equity and efficiency: The initial distribution of property rights or emission permits determines whether a competitive global CO_2 permit market will operate efficiently.[3] Prior to now it has been generally assumed that the manner in which emission permits are initially distributed will not affect the efficiency of the market.[4] We show here that of all the many possible ways of distributing a given total of emission rights, very few are compatible with efficient markets. In this case equity and efficiency are not orthogonal, as in the first and second theorems of welfare economics for standard competitive markets. How does this happen?

The key to this result is the fact that the atmospheric concentration of CO_2 is a privately produced public good, privately produced but affecting the utility levels of all people. The reason is that CO_2 mixes thoroughly in the atmosphere, leading to a uniform concentration over the globe. Therefore, we have a global public good. People or regions cannot choose their concentration levels independently. However, the concentration is determined by every individual who runs a car or a heating furnace and by every firm operating transportation or burning fuel in any other way.[5] Therefore, we have a privately produced public good. The fact that CO_2 concentration is a privately produced public good affecting the welfare levels of individuals leads to the equity-efficiency interaction. As noted, everyone has de facto to consume the same CO_2 concentration. For efficiency this common level must be what they demand, given prices and their incomes. In summary, for agents to demand freely the same amounts of CO_2 at an equilibrium requires a particular choice of the distribution of income.

Similar points were made in Chichilnisky [2] and Chichilnisky and Heal [4],[6] where this simple observation was shown to have other far-reaching consequences. In particular these papers establish that the equalization of marginal abatement costs across countries is neither sufficient nor necessary for Pareto

[3] The term *efficiently* here is used in the standard economic sense of "so as to attain Pareto efficiency."

[4] It will of course affect the distribution of income resulting from the operation of the market. This is the original Coase [9] position: that whatever the initial distribution of permits, trading rights can bring about a Pareto-efficient allocation of resources. In fact a stronger claim is sometimes made: that the equilibrium allocation of resources is not affected by the initial distribution of permits. Clearly, the conditions for this stronger claim to be true are very restrictive indeed—a total absence of income effects; see Milgrom and Roberts [22], chapter 2.

[5] Carbon dioxide, a public bad, is a by-product of the consumption and production of private goods.

[6] There is also an early discussion of closely related issues in Laffont [19] and Eyckmans et al. [13].

efficiency: Pareto-efficient allocations may have different marginal costs. Here we show that this line of argument, when developed further, implies that efficiency and distribution cannot be separated in environmental markets. Efficiency requires an appropriate distribution of property rights. The fact that many distributions of property rights lead to inefficient outcomes allows us to construct an example of a two-region world in which a transfer of property rights from the North to the South, accompanied by a decrease in the total of emission permits, leaves both regions better off.

Finally, we investigate the extent to which an equilibrium concept related to that of Lindahl is the appropriate concept in permit markets. There is a simple reason that this might be so: A Lindahl equilibrium is the only market equilibrium known to lead to Pareto efficiency with public goods.[7] As a permit market is a market that determines the production of public goods, we might therefore expect that efficiency would require the key feature of a Lindahl equilibrium, namely, a multiplicity of prices, in fact one price per pair of traders. In a Lindahl equilibrium each producer of a public good is paid for her production by each consumer, and the per unit payment typically varies from consumer to consumer. Therefore, relative to the framework of a Lindahl equilibrium, a permit market as formalized here is an "incomplete market" because everyone pays the same price for the permits. This can be interpreted as assuming that the "individualized" markets between buyers and sellers are missing. Our main result shows that, in a certain sense, it is possible to compensate for the absence of individualized markets by reallocating property rights in tradable permits.[8]

3.2 Efficiency and International Emissions

Following the model set out in Chichilnisky [2] and developed further in Chichilnisky and Heal [4], we consider a world economy with I regions, $I \geq 2$, indexed by $i = 1, ..., I$. Each region has a utility function u_i, which depends on its consumption of a vector of private goods $c_i = (c_{i,1}, c_{i,2}, ..., c_{i,M})$, where M is the number of private goods (indexed by m), and also on the quality of the world's atmosphere, a, which is a public good.[9] The quality of the atmosphere a can be thought of as a measure of abatement. It could be measured by, for example, the reciprocal or the negative of the concentration of CO_2: The more abatement there is, the lower is this concentration. The concentration of CO_2

[7]See Foley [14].

[8]The dimensionality of the space of permit allocations equals that of the space of Lindahl prices needed to complete the market, so that the two approaches are mathematically equivalent.

[9]Formally, $u_i(c_i, a)$ measures welfare, where $u_i\colon \Re^{M+1} \to \Re$ is a continuous, strictly concave and increasing function. It is assumed to be twice continuously differentiable.

is "produced" by emissions of carbon, which are positively associated with the levels of production of private goods. Let y_i be a vector $(y_{i,m})$ in R^M giving the production levels of the M private goods in country i. Then

$$a = \sum_{i=1}^{I} a_i,\ a_i = \Phi_i(y_i) \quad \text{for each country } i = 1, \ldots, I, \quad \text{and}$$

$$\frac{\partial \Phi_i}{\partial y_{i,m}} < 0 \qquad \forall i \quad (3.1)$$

The production functions or abatement functions Φ_i are continuously differentiable and strictly concave and show the trade-off between the level of abatement or quality of the atmosphere and the output of consumption.[10] An allocation of consumption and abatement across all countries is a vector

$$(c_1, a_1, \ldots, c_I, a_I) \in \Re^{(M+1)I},$$

as for each of the I regions there are M private goods and one level of abatement. An allocation is feasible if it satisfies constraint (3.1), and the condition that the total consumption of each private good worldwide be equal to the total production, that is,

$$\sum_{i=1,\ldots,I} c_i = \sum_{i=1,\ldots,I} y_i \qquad (3.2)$$

Constraint (3.2) allows private goods to be transferred freely between regions; that is, it allows unrestricted lump-sum international redistributions. This is a rather strong assumption that gives a full first-best solution. It is not of course equivalent to modeling free trade in international markets because the latter requires that each region trade within its budget: each region must satisfy a balance of payments condition.[11]

3.2.1 Characterization of Pareto Efficiency — In this section we provide a characterization of Pareto-efficient allocations. This section does not address

---

[10] We can suppose that the functions Φ_i embody information about countries' initial endowments of goods. By assuming strict concavity, we are bypassing the possible nonconvexities associated with externalities (Starrett [24]).

[11] See Chichilnisky and Heal [5]. International trade between regions would require that

$$\forall i, \qquad (c_i - y_i)p = 0, \qquad (3.3)$$

where $p \in \Re^m$ is a world price vector. This condition requires the value of the difference between consumption and production to be zero at world prices, which implies that for each region the value of goods that are imported and for which consumption exceeds production equals the value of goods that are exported and for which production therefore exceeds consumption.

any institutional framework, as it does not presume any structure, such as emission markets or emission taxes. It describes the conditions that any resource allocation must satisfy if it is efficient, whatever the institutional structure through which it is implemented.

Lump-Sum Transfers

An allocation is called *feasible with lump sum transfers* if it satisfies constraints (3.1) and (3.2). Such an allocation $(c_1^*, a_1^*, ..., c_I^*, a_I^*) \in R^{(M+1)I}$ is *Pareto efficient* if there is no other feasible allocation at which every region's utility is at least as high, and one's utility is strictly higher.[12] It is immediate therefore that a Pareto-efficient allocation solves the following problem:

$$\max u_i(c_i, a) \quad \text{subject to} \quad u_k(c_k, a) = N_k \ \forall k \neq i, \ k = 1, ..., I,$$
$$\textstyle\sum_{i=1}^{I} y_{i,m} = \sum_{i=1}^{I} c_{i,m} \ \forall m,$$
$$a_i = \Phi_i(y_i), \quad \text{and} \quad \textstyle\sum_i a_i = a. \tag{3.4}$$

Here N_k is a utility level specified for region k.[13]

To solve problem (3.4) we can write out the corresponding Lagrangian

$$L = u_i\left(c_i, \sum_{i=1}^{I} \Phi_i(y_i)\right) + \sum_{k=1,...,I,k\neq i} \lambda_k\left(u_k\left(c_k, \sum_{i=1}^{I} \Phi_i(y_i)\right) - N_k\right)$$
$$+ \sum_{m=1}^{M} \theta_m\left(\sum_{i=1}^{I} y_{i,m} - \sum_i c_{i,m}\right),$$

where a has been replaced by $\Sigma_i \Phi_i(y_i)$ in view of (3.1). Differentiating L with respect to the components of c_i and y_i and equating to zero gives the first-order conditions for efficiency (3.5) and (3.6):

$$\overbrace{\frac{\partial u_i}{\partial c_{i,m}} = \lambda_k \frac{\partial u_k}{\partial c_{k,m}} \quad \forall m = 1, ..., M \text{ and } \forall k \neq i,}^{\text{equal marginal valuations of consumption}} \tag{3.5}$$

[12] A Pareto-efficient allocation can be characterized as a solution to the problem of maximizing the utility of a designated region, subject to the others all reaching prescribed utility levels. The solutions of this problem (as the prescribed utility levels vary over all feasible values) describe the utility possibility frontier.

[13] Observe that the second line of this problem allows unrestricted international lump-sum redistribution. Worldwide consumption has to equal worldwide production, with no region-by-region balanced budgets required.

where i is the designated region whose utility is being maximized, λ_k is a Lagrange multiplier associated with the constraint that region k should reach a specified welfare level, and

$$\overbrace{\frac{\partial \Phi_i}{\partial y_{i,m}} = \frac{-\dfrac{\partial u_i}{\partial c_{i,m}}}{\Sigma_k \lambda_k \dfrac{\partial u_k}{\partial a}} \ \forall m, \quad \text{and for } k \neq i, \quad \frac{\partial \Phi_k}{\partial y_{k,m}} = \frac{-\lambda_k \dfrac{\partial u_k}{\partial c_{k,m}}}{\Sigma_k \lambda_k \dfrac{\partial u_k}{\partial a}} \ \forall m.}^{\text{Lindahl-Bowen-Samuelson condition}} \tag{3.6}$$

Each of these systems of equations has a simple interpretation. The first system, (3.5), requires that for any good m the marginal social value of consumption be the same for all regions i. We refer here to the "marginal social value of consumption by region i" because the marginal utilities of consumption are weighted by the terms λ_k, which represent the shadow price or social value of utility in region k. The second set of equations, (3.6), is a slight modification of the conventional Lindahl-Bowen condition, popularized by Samuelson. It requires that the marginal rate of transformation between the public good and a private good be equal to the sum of the marginal rates of substitution. (See also chapter 13 for a detailed analysis of efficiency conditions.)

Without Lump-Sum Transfers

If we restrict international lump-sum redistributions, the corresponding characterization of (constrained) Pareto efficiency is different. For example, if we model an autarchic world where in each region consumption is required to equal production, the second line of the problem (3.4) is dropped and the vector y_i in the third line replaced by c_i. In this case the necessary conditions for Pareto efficiency are just (3.6). Condition (3.5) is no longer required.

Should Marginal Costs Be Equal?

Note that the marginal cost of abatement in region i in terms of good m is just the reciprocal of the marginal productivity with respect to m of the function Φ_i:

$$MC_{i,m}(a_i) = -\frac{1}{\dfrac{\partial \Phi_i}{\partial y_{i,m}}}. \tag{3.7}$$

PROPOSITION 1[14] At a Pareto-efficient allocation $(c_1^*, a_1^*, ..., c_I^*, a_I^*)$, in each country the marginal cost of abatement $MC_i(a_i^*)$ in terms of private good m is inversely proportional to the marginal valuation of the private good m, $\lambda_i \partial u_i / \partial c_{i,m}$. In particular, at a Pareto efficient allocation, the marginal costs will be equal across countries if and only if the marginal valuations of the private goods are equal; that is, for each good m, $\lambda_i \partial u_i / \partial c_{i,m}$ is independent of i.

It follows that with lump-sum transfers, as represented by constraint (3.2), marginal costs will always be equalized, as private goods can always be shifted between countries by lump-sum redistributions to equate their marginal valuations. However, if each country is required to consume what it produces or is required to trade internationally subject to a standard balance of trade constraint, this is not true.[15] Therefore, in general equalization of marginal costs across countries is not necessary for efficiency.

3.3 International Emission Markets

In section 3.2 we characterized in equations (3.5) and (3.6) allocations that are Pareto efficient in an institution-free framework as well as those in which each region consumes what it produces.

Next we introduce an institutional framework: an international market for tradable permits. The aim is to investigate the first-best efficiency of the equilibria in this market. To model a policy-relevant situation, assume that the initial distribution of emission permits is the only variable used to address distributional issues.[16] Each region is given an initial endowment of permits to emit E_i units of CO_2, where $\Sigma_i E_i = E^*$, the desired level of total emissions. Regions trade these and behave as price takers in a market in which there is a single price p_e for a permit to emit one unit.

If the number of units of CO_2 emitted exceeds the number of permits a region has, the region must buy the difference in the permit market. Otherwise, it can sell excess permits and use the proceeds to buy private goods at prices p_l. A region therefore maximizes its utility $u_i(c_i, a)$ subject to the following budget constraint:

$$\sum_{m=1}^{M} c_{i,m} p_m = \sum_{m=1}^{M} y_{i,m} p_m - p_e(E_i + a_i). \tag{3.8}$$

[14]Chichilnisky and Heal [4] established the following proposition in the case of one private good. The extension to the present case, which differs only in having many private goods is immediate.

[15]See Chichilnisky and Heal [5].

[16]In particular, unrestricted lump-sum redistributions of private goods are not possible.

The difference between actual emissions e_i and target emissions E_i is $e_i - E_i = e_i^N - a_i - E_i$, where e_i^N is the emission level of region i when abatement is zero.[17] The budget constraint requires that in each region the value of consumption equal the value of production plus the net revenue from the sale of permits. This can be rewritten as

$$\left(\sum_{m=1}^{M} c_{i,m} p_m - \sum_{m=1}^{M} y_{i,m} p_m\right) = -p_e(E_i + a_i). \tag{3.9}$$

The left-hand side is the difference between the value of domestic consumption and production, that is, the balance of trade. A surplus of consumption over production[18] is funded by the revenue generated by sales of permits in international markets. Conversely, a net purchase of permits in international markets has to be matched by a surplus of production over consumption and therefore a net export position.

A comparison of the balance-of-trade condition (3.9) with the actual budget constraint (3.3) suggests that controlling the initial endowments of emission rights can act as a substitute for lump-sum transfers. This point is developed later in section 3.4.

Each region seeks to maximize its utility $u_i(c_i, a)$ subject to the budget constraint (3.8) and to the production relations given in (3.1). We assume that in so doing it supposes the total level of emissions to be fixed at E^*, the desired total level. This in effect implies the existence of a credible intergovernment agency (the UNFCC, for example) that sets and implements global emission targets.[19]

3.3.1 Market Behavior — Maximizing its welfare subject to the budget constraint (3.9), each region chooses consumption levels and abatement or emission levels to satisfy the following first-order conditions:

$$\text{MRS} = \text{price ratio, or} \quad \frac{\dfrac{\partial u_i}{\partial c_{i,l}}}{\dfrac{\partial u_i}{\partial c_{i,j}}} = \frac{p_l}{p_j}, \tag{3.10}$$

---

[17] For simplicity we have dropped the constant terms in e_i^N.

[18] That is, a position of net imports.

[19] An alternative, which we do not explore here, would be to look for a Nash equilibrium in countries' abatement levels. In this Nash case each country would observe the emissions of each other and then choose its optimal emission level on the assumption that these levels are fixed. This approach is developed in Heal and Lin [18] (chapter 5 in this volume). For a similar development, see Dasgupta and Heal [12], chapter 3.

and

$$\text{MRT} = \text{price ratio, or} \quad \frac{\partial \Phi_i}{\partial y_{i,l}} = -\frac{p_l}{p_e}. \tag{3.11}$$

These are standard conditions for utility maximization subject to production and budget constraints. First-order condition (3.10) just requires that marginal rates of substitution between goods be equated to their price ratios, and (3.11) requires tangency between the production possibility frontier and an isoprofit hyperplane.

3.3.2 Market Solutions that Are Not Pareto Optimal — How do first-order conditions (3.10) and (3.11) characterizing a region's optimal market choice compare with conditions (3.5) and (3.6), which describe Pareto-efficient allocations? Condition (3.11) from regional utility maximization is the same as the Bowen-Lindahl-Samuelson condition (3.6) for the efficient provision of public goods, provided that

$$\frac{p_m}{p_e} = \frac{\dfrac{\partial u_i}{\partial c_{i,m}}}{\Sigma_{k=1}^{I} \lambda_k \dfrac{\partial u_k}{\partial a}} = \frac{\lambda_k \dfrac{\partial u_k}{c_{k,m}}}{\Sigma_{k=1}^{I} \lambda_k \dfrac{\partial u_k}{\partial a}} \quad \forall k \neq i. \tag{3.12}$$

Condition (3.12) can hold only if the marginal valuations of the *m*th private good, $\partial u_i / c_{k,m}$ and $\lambda_k (\partial u_k / \partial c_{k,m})$, are independent of i and k, that is, are the same for all regions.

Condition (3.5) is required for Pareto efficiency—equalization of the marginal valuation of consumption across countries—and automatically implies this. However, there is nothing equivalent to (3.5) in the solutions to the regions' optimization problems. The only other condition from each regions' own optimization problems is (3.10), which does not imply equality of marginal valuations across countries.

In brief, utility maximization subject to the budget constraint (3.8) does not lead to the conditions needed for Pareto efficiency, as illustrated in figures 3.1 and 3.3 below. The next section provides a simple geometric example illustrating this result. There is an additional requirement represented by (3.5). For the Bowen-Lindahl-Samuelson condition to hold, we need the marginal valuation of consumption to be the same in all regions; that is, $\partial u_i / \partial c_{i,m} = \lambda_k (\partial u_k / \partial c_{k,m}) \; \forall m, \forall k \neq i$. This condition would of course be satisfied if there were policy instruments available to redistribute freely all resources without restriction across regions—if, for example, lump-sum redistributions were possible. In the absence of such instruments, what is required to ensure that (3.5) is met and efficiency attained in the permit market?

3.4 Equity and Efficiency in Permit Markets

Competitive permit markets do not generally lead to the conditions for Pareto efficiency because there is nothing that ensures that condition (3.5), $\partial u_i / \partial c_{i,m} = \lambda_k (\partial u_k / \partial c_{k,m})$ $\forall m = 1, \ldots, M$ and $\forall k \neq i$, is satisfied. Now this is clearly a condition on the distribution of income or wealth. Look in more detail at the determinants of the terms $\partial u_i / \partial c_{i,m}$. As $u_i = u_i(c_i, E^*)$, where E^* is fixed, the derivatives of u_i with respect to consumption can depend only on consumption levels.[20] In the absence of policy instruments to achieve unrestricted redistributions across regions, the only variables then available for ensuring that marginal social valuations of consumption are equalized are the initial allocations of permits, and therefore only those initial permit allocations that ensure that (3.5) is satisfied will lead to Pareto-efficient allocations. We formalize this in the following and show that very few initial allocations satisfy this condition.

3.4.1 Why Distribution Matters — An intuitive explanation for the dependence of efficiency on distribution is as follows. Because we are trading a public good, everyone must consume the same amount at equilibrium, a physical requirement resulting from the fact that the gas CO_2 distributes uniformly across the world. Achieving more targets typically requires more instruments, and here the extra instruments are the distribution of emission permits or property rights. The efficient distributions of property rights are those at which there are market-clearing prices such that all regions demand freely the same level of the public good. If regions' preferences were similar, this would require similar income levels. A useful comparison is with a Lindahl equilibrium, the standard market equilibrium concept for public goods, in which the extra instruments are provided by region-specific prices. Recall that at a Lindahl equilibrium the prices for public goods will typically be different for different consumers, so that with Lindahl markets different regions would pay different prices for emission permits. In this case permit trading would not equalize marginal abatement costs across regions.

[20]These in turn depend, by the budget constraint (3.8), on prices p_m, production levels $y_{i,m}$, abatement levels a_i, and initial endowments of emission rights E_i. Once prices are given, production and abatement levels are fully determined by (3.11).

Another explanation for the significance of the distribution of property rights is as follows:

1. Trading emission permits naturally leads to the equalization of marginal abatement costs across countries. By obvious arguments each country equates the marginal cost of abatement to the price of an emission permit, which by assumption is the same for all countries (see equation [3.11]).
2. Equalization of marginal costs is efficient only if marginal social valuations of consumption are equalized (see proposition 1). Therefore, permit trading is efficient only if marginal social valuations of consumption are equalized. This can be achieved only by an appropriate redistribution of wealth.
3. The assignment of property rights brings about a redistribution of wealth. The efficient allocations of permits are those that equate marginal valuations of consumption.

3.4.2 An Example: One Private Good and Two Regions — Imagine two regions trading one private good and one public good (abatement). Figure 3.1 shows the abatement-production frontier and the preferences over combinations of public and private goods for each region. An emission level E^* has been chosen that we assume is a level associated with a Pareto-efficient allocation. Therefore, the question before us is, When can we attain this efficient allocation of resources by trading emission permits?

The total abatement level of the two regions must be $-E^*$, and because they are identical, each must produce a level of abatement of $-E^*/2$. Each region's production of the private good is now determined to be the level that corresponds to an abatement level of $-E^*/2$, so that the production points of the regions are now determined as in figure 3.1. As a result, the relative price of the public and private good is determined and is the slope of the frontier at this point. Each region's consumption of the public good abatement is the total amount of abatement produced, $A^* = -E^*$, and its consumption of the private good is determined by maximizing utility subject to the equation

$$c_i = y_i - p_e(E_i + a_i),$$

where c_i and y_i are region i's consumption and production of the single private good and p_e is the relative price of the emission permits. Here y_i, p_e, and a_i are fully determined from the total level of emissions E^* by the following chain. Total emissions E^* imply individual emissions $E^*/2$, which imply abatement levels, which imply production levels and the price of permits relative to the consumption good. Therefore, only E_i, the initial endowment of permits, is

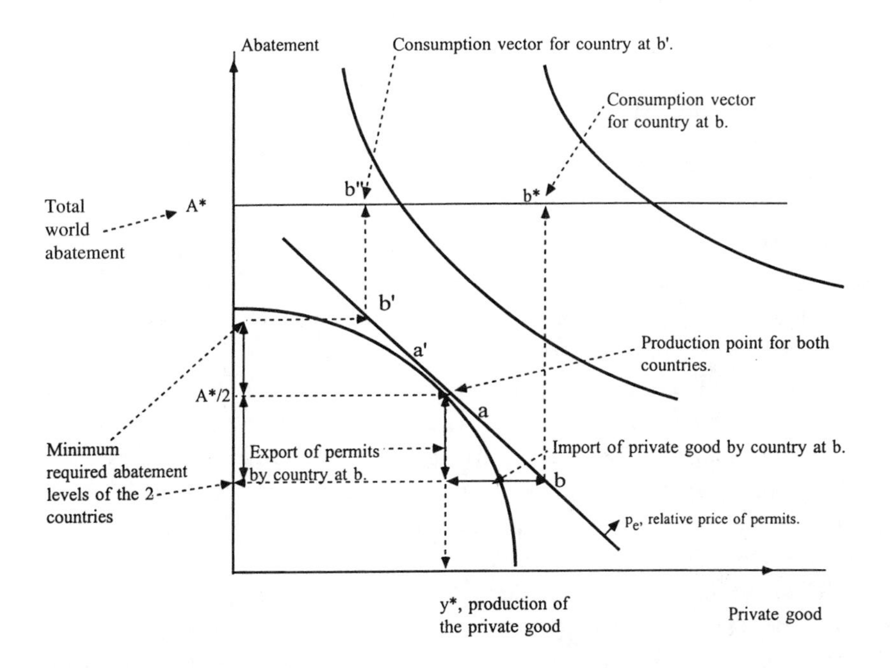

FIGURE 3.1 Only specific distributions of property rights lead to Pareto efficiency.

available to control consumption c_i. This variable must therefore be used to ensure that marginal valuations of the private good satisfy the condition (3.5) needed for Pareto efficiency. Figure 3.1 illustrates how this can be done. If both regions are given endowments of permits equal to their levels of emission, neither will trade permits, and each will consume the amount of the private good that it produces. They will consume levels of the private good given by the horizontal coordinate of the production point in figure 3.1, y^*. Their consumption of the public good abatement will be the sum of the production levels of both regions, A^*. Each region's consumption vector has a vertical coordinate equal to A^* and a horizontal coordinate equal to its consumption of the private good, namely (y^*, A^*).[21]

Consider further the case in which both regions have an initial allocation of permits equal to their production of CO_2. As they both neither import nor export the private good and so consume and produce the same amounts and also

[21] In general this is production plus imports from the sale of permits or minus exports to pay for the purchase of permits. Both are zero if countries are given endowments of permits equal to their levels of emission.

consume the same amount of the public good, their marginal valuations of the private goods must be the same.

Suppose now that condition (3.5) requires for efficiency that the marginal valuations of the private good are different, that is, that $\partial u_1/\partial c_{1,l} = \lambda_2 (\partial u_2/\partial c_{2,l}) < \partial u_2/\partial c_{2,l}$, where 1 and 2 are the two regions and l denotes the single private good. Then to satisfy (3.5) region 2's consumption of the private good has to be decreased and region 1's increased from their common production level. This can be achieved by giving region 1 an endowment of permits (b) in excess of its emissions and region 2 an endowment (b′) less than its emissions. Region 1 then increases its consumption of the private good by selling its spare permits and using the proceeds to buy the private good, whereas region 2 is forced to sell the private good to buy permits. Region 1's marginal utility of the private good will be less than region 2's, and the ratio will decrease continuously from unity as region 1's initial endowment of permits is raised above the emission level corresponding to its production of the private good (and region 2's is correspondingly reduced).

Consider the straight line p_e through the regions' production points tangent to the production frontier, as shown in figure 3.1. Each region produces a mix of abatement and private good given by the point of tangency and then trades private goods for emission permits along the line tangent to the production frontier. If it has more permits than needed (i.e., more than $E^*/2$), it will add consumption of the private good by selling permits and buying the private good along the tangency line, whose slope is the relative price of permits and the private good. As it moves along this line, its consumption of abatement remains constant.[22] However, its consumption of the private good changes. The other region will be symmetrically placed on this line relative to the production point (y^*, A^*). In this way we can reach an allocation at which all markets will clear, total emissions will be E^*, and condition (3.5) needed for efficiency will be satisfied. We can do this by picking the permit allocations and therefore consumption levels of the private good correctly. As the ratio of the regions' marginal utilities changes continuously with their initial allocations of permits, there will generally be at most a finite number of initial allocations at which the efficiency conditions hold. In this simple example, there will be just one initial distribution of permits that will lead to efficiency. This argument illustrates the following result.

PROPOSITION 2 Let E^* be the level of total emissions at a Pareto-efficient allocation of resources in the economy described in section 2 with one private

[22] It is selling surplus permits, not abatement.

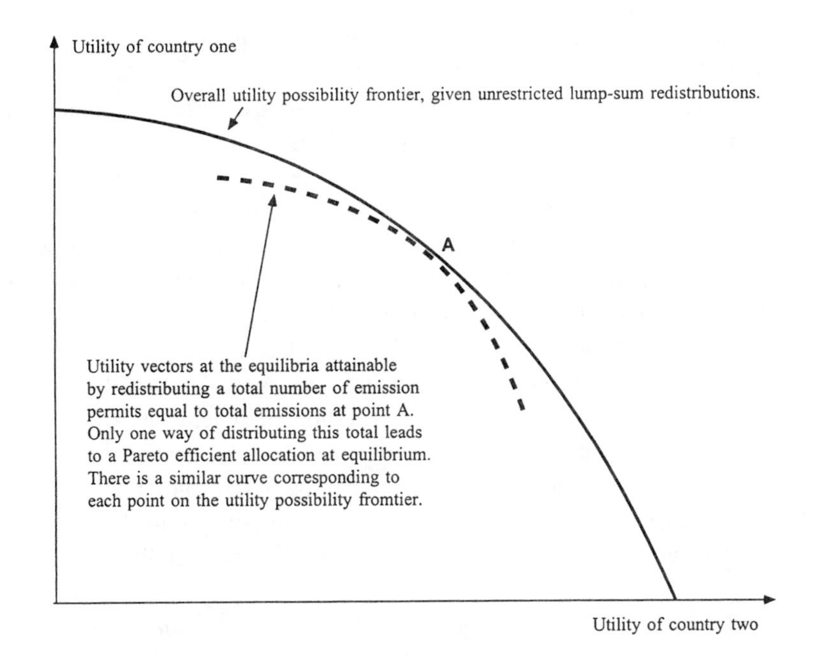

FIGURE 3.2 Redistribution of a fixed total of emission rights leads to a utility possibility curve inside the Pareto frontier.

good and two regions. Then of all possible ways of allocating the total emission E^* among the regions as initial endowments, only a subset of measure zero will lead to market equilibria that are Pareto efficient. Alternatively, almost every allocation of permits between regions will lead to inefficient outcomes.

For a proof, see the Appendix.[23]

The diagrammatic analysis illustrating proposition 2 can in fact be pushed further, as in figure 3.2. As figure 3.1 shows, each possible distribution of the total emission permits E^* between the two regions leads them to a pair of levels of consumption of the private good given by the horizontal coordinates of pairs of points, such as (a, a') or (b, b'), which are symmetrically placed on the line that is tangent to the production frontier at the production point. These pairs of points in turn give rise to consumption vectors for the public and private and

[23]The results in proposition 2 are robust. They hold not only for first-best, or Pareto, efficiency, as discussed previously, but also for efficiency subject to an arbitrary abatement constraint (see Heal [17]). In this case it is still true that only certain specific distributions of emission rights are compatible with efficiency, defined now as maximization of the sum of utilities subject to feasibility constraints and also to a politically imposed constraint on the level of emissions.

private goods, together represented by points such as b'' and b^* in figure 3.1. From figure 3.1 we can ascertain the utility levels of these points. Suppose that we plot the utility levels arising from all such possible distributions of the total E^* permits. What does this set of points look like?

We know that few points will be Pareto efficient, so that this must form a curve largely inside the utility possibility frontier, touching this frontier at a finite number of points, at most. In fact in the present two-region fully symmetric case, it is easy to see that once we have an allocation of permits that satisfies (3.5), departures from this allocation increase the difference from equality of the two sides in (3.5), so that the efficient allocation is unique. Figure 3.2 therefore illustrates the set of utility vectors associated with different allocations of the total of E^* permits and also shows the overall utility possibility frontier. Each point on the frontier corresponds to a different total emission level and therefore to a different total number of permits, and for each point on the frontier there is one way of allocating the corresponding total of permits that is efficient and gives the utility vector on the utility possibility frontier. [24]

3.4.3 Pareto-Improving Reallocations from North to South: Win-Win Solutions — A consequence of proposition 2 is that in general a competitive market in emission permits admits changes in the total and the distribution of permits that are Pareto improving, something that is of course not possible in competitive markets for private goods. Figure 3.3 illustrates such a situation.

This figure refers to two regions, called, for obvious reasons, North and South. Both are identical in production possibilities and preferences. The production frontier and two indifference curves are shown. We consider a decrease in the total number of emission permits (an increase in abatement) coupled with a transfer of permits from the North to the South and show that this can be Pareto improving for both regions simultaneously.

The initial abatement level is given by the vertical coordinate of the lower of the two solid horizontal lines and the final by that of the higher. The initial production point is therefore determined so that abatement by each region is half the initial total. Relative prices of permits and the private good are given by the slope of the production frontier at this point, and the initial permit distribution is such that the initial abatement levels of the North and South are as shown. This leads to consumption levels for the North and the South on the higher and the lower indifference curve, respectively.

Now consider a different and lower total of emission permits, one corre-

[24] Lin [20] solves analytically for the curves in figure 2 for specific utility and production functions.

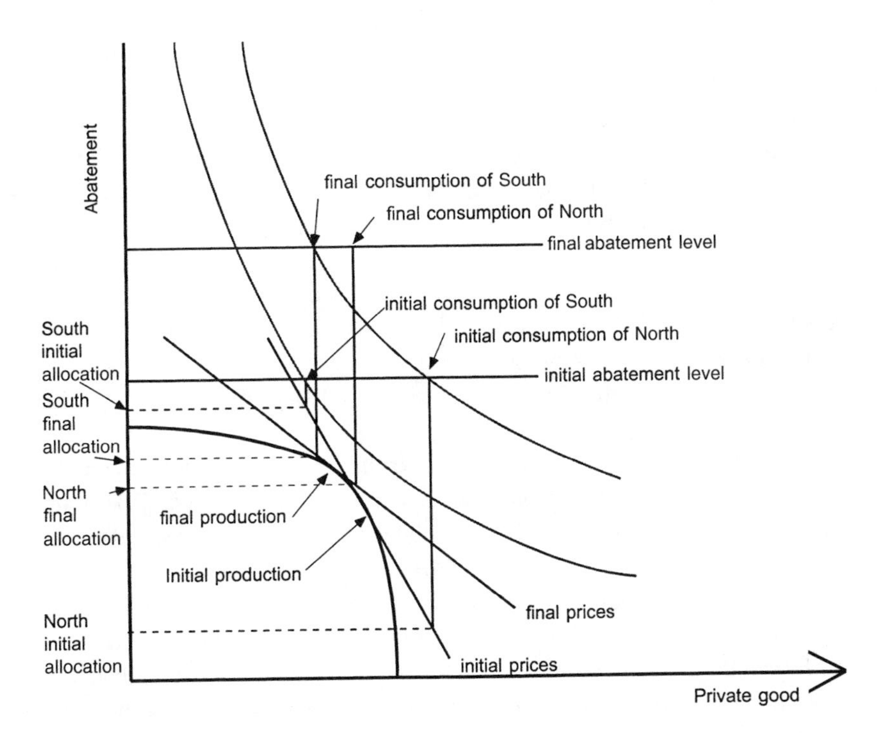

FIGURE 3.3 A redistribution of property rights from North to South can make both better off.

sponding to the higher final abatement level. Each region has to produce less of the private good and abate more, as shown by the point "final production." At the same time as the total abatement target is raised, the South's abatement target is lowered (from "South initial" to "South final"), and the North's is raised. In other words permits are transferred from North to South while the total is reduced. The new equilibrium consumption levels are as shown. Both regions are now better off, and the level of world emissions is lower.

3.4.4 The General Case — The result in proposition 2 holds for the general case, but the argument is less intuitive. Formally, we establish the following proposition:

PROPOSITION 3 Let E^* be the level of total emissions at a Pareto-efficient allocation of resources in the economy described in section 2. Assume that regions maximize utility subject to the budget constraint (3.8) given by the ability to trade emission permits. Assume furthermore that a regularity condition defined in the Appendix is satisfied. Then of all possible ways of allocating the total emission E^* among the regions as initial endowments, only a subset of measure zero will lead to market equilibria that are Pareto efficient. Alternatively, almost every allocation of permits between regions will lead to inefficient outcomes. If the inequality $(I - 1) + m \leq (I - 1) \times m$ holds, then only a finite number of ways of allocating the emission rights lead to efficiency.

The proof of this proposition is given in the Appendix. Strict concavity and the regularity assumption are needed for this result. Otherwise, one can construct counterexamples. For example, with quasi-linear preferences of the form $u_i(a) + \alpha_i c_i, \alpha_i > 0$, there might be infinitely many allocations of permits that will lead to efficient outcomes.

Although the dependence of efficiency on distribution runs quite counter to the thrust of the first and second welfare theorems, there are parallels in the literature. For example, in economies with increasing returns to scale, there are some allocations of a given total of initial endowments that are compatible with attainment of efficiency at a marginal cost-pricing equilibrium and some that are not (see Brown and Heal [1]). The orthogonality of efficiency and distribution might therefore be limited to "classical" economic environments free from increasing returns and public goods or externalities. In fact, there is a perspective from which increasing returns and public goods are closely related, so that this connection is not surprising.

3.5 Lindahl Permit Markets

In this section we compare the permits markets modeled previously in which there is a uniform price for all buyers and sellers, with a Lindahl-type framework in which each region may pay a different price for emission permits. This is motivated by reference to a Lindahl equilibrium, at which each producer of a public good is paid by every consumer for each unit produced, and in principle all consumers may pay different prices to a given producer. [25] In the present context the exact analog would be the following. Any region considering producing one more unit of emissions would have to purchase from every other the right to emit that extra unit. It would therefore have to buy an emission permit from each affected region, with possibly a different price ruling in each bilateral trade. This would give as many prices as there are in a Lindahl equilibrium.

[25] For a definition of Lindahl equilibria, see Foley [14] or Dasgupta and Heal [12].

An alternative way of interpreting such a model is to think of markets for externalities, as described by Meade [21] in his famous bees and apples example (see Dasgupta and Heal [12] for an exposition relevant to the present model). In this context each pairwise externality is a separate commodity, separately priced. There are therefore as many prices as there are pairs of interacting producers and consumers of externalities. In the present context, as the externalities imposed on a region depend only on the sum of emissions by other and not on the identities of the emitters, the dimensionality can be reduced so that the number of prices equals the number of regions rather than the number of pairs. There is a price for buying the right to pollute from each region that is the same for every buyer. At a normal Lindahl equilibrium, there are I^2 prices, one between each pair of the I regions, as each is both a buyer and a seller of emission rights, whereas with each charging a different price for a permit, there are only I prices. By comparison, in the framework modeled previously, there is only one price.

If each region faces a region-specific price for emission permits, the budget constraint (3.8) is changed to

$$\sum_l c_{i,l} p_l = \sum_l y_{i,l} p_l - p_{i,e}(E_i + a_i), \tag{3.13}$$

where $p_{i,e}$ is the price of an emission permit to i. Instead of (3.11), each region's first-order condition in production now becomes

$$\frac{\partial \Phi_i}{\partial y_{i,l}} = -\frac{p_l}{p_{i,e}}. \tag{3.14}$$

Recall that a necessary condition for efficiency is (3.6):

$$\overbrace{\frac{\partial \Phi_i}{\partial y_{i,l}} = \frac{-\dfrac{\partial u_i}{\partial c_{i,l}}}{\Sigma_k \lambda_k \dfrac{\partial u_k}{\partial a}} \; \forall l, \quad \text{and for } k \neq i, \quad \frac{\partial \Phi_k}{\partial y_{k,l}} = \frac{-\lambda_k \dfrac{\partial u_k}{\partial c_{k,l}}}{\Sigma_k \lambda_k \dfrac{\partial u_k}{\partial a}} \; \forall l,}^{\text{Lindahl-Bowen-Samuelson condition}}$$

so that in place of (3.12) the condition for permit markets to attain efficiency is

$$\frac{p_l}{p_{k,e}} = \frac{\lambda_k \dfrac{\partial u_k}{\partial c_{k,l}}}{\Sigma_k \lambda_k \dfrac{\partial u_k}{\partial a}} \; \forall k. \tag{3.15}$$

Because the permit price $p_{k,e}$ is region specific, this condition can now be satisfied without $\lambda_k (\partial u_k / \partial c_{k,l})$ being the same for all k. In other words this condition for Pareto efficiency can be satisfied now without an optimal distribution of income or wealth, which equates marginal valuations of consumption. Therefore, if redistribution of private goods or emission permits is ruled out, there is a real efficiency gain to having permit prices that are region specific, for without them it would not be possible to attain a Pareto-efficient allocation.

Appendix

Proof of Proposition 3

The first-order conditions for efficiency are

$$\overbrace{\frac{\partial u_i}{\partial c_{i,l}} = \lambda_k \frac{\partial u_k}{\partial c_{i,l}} \qquad \forall l = 1, \ldots, m \quad \text{and} \quad \forall k \neq i,}^{\text{equal marginal valuations of consumption}}$$

where region i is the designated region whose utility is being maximized, λ_k is a Lagrange multiplier associated with the constraint that region k reach a specified welfare level, and

$$\overbrace{\frac{\partial \Phi_i}{\partial y_{i,l}} = \frac{-\dfrac{\partial u_i}{\partial c_{i,l}}}{\Sigma_k \lambda_k \dfrac{\partial u_k}{\partial a}} \; \forall l, \quad \text{and for } k \neq i, \quad \frac{\partial \Phi_k}{\partial y_{k,l}} = \frac{-\lambda_k \dfrac{\partial u_k}{\partial c_{k,l}}}{\Sigma_k \lambda_k \dfrac{\partial u_k}{\partial a}} \; \forall l.}^{\text{Lindahl-Bowen-Samuelson condition}}$$

The first set of conditions, $\partial u_i / \partial c_{i,l} = \lambda_k (\partial u_k / \partial c_{k,l}) \; \forall l, \forall k \neq i$, constitute a system of $(I - 1) \times m$ equations. If they are satisfied, then the second set of conditions is also satisfied at an equilibrium of a permit market. Therefore, we need to check only when the equal marginal valuation conditions are satisfied. Rewrite them as

$$\frac{\partial u_i}{\partial c_{i,l}} - \lambda_k \frac{\partial u_k}{\partial c_{k,l}} = 0. \tag{3.16}$$

Efficiency now requires that we locate a zero of a system of $(I - 1) \times m$ nonlinear equations given by (3.16).

What are the independent arguments of the functions in (3.16)? Note that once the prices of all goods are chosen, the production levels of private goods and of abatement are determined by equation (3.11), giving first-order conditions in production. And these levels, together with prices and endowments of permits, determine consumption levels through the budget constraint (3.8) and the first-order conditions on consumption (3.10). Therefore, the arguments of (3.16) can be taken to be E_i, $i = 1, ..., I$ and p_l, $l = 1, ..., m$ and e. Now, as the E_i are nonnegative and sum to a fixed number, they form a space of dimension $(I - 1)$. As there are only m relative prices, the left hand side of system (3.16) is a function, call it Ω, defined on $\Re^{(I-1)} \times \Re^m = \Re^{(I-1)+m}$. In fact it is defined on a subset of $\Re^{(I-1)+m}$ because if E is the vector of endowments and p the vector of relative prices, then $p = p(E)$: Equilibrium relative prices are determined by initial endowments. The graph of $p = p(E)$ is a subset of $\Re^{(I-1)+m}$ and indeed would be the equilibrium manifold of the economy under suitable regularity conditions.

The function Ω takes values in $\Re^{(I-1)\times m}$:

$$\Omega\colon \Re^{(I-1)+m} \to \Re^{(I-1)\times m},\ \Omega(x) = \frac{\partial u_i(x)}{\partial c_{i,l}} - \lambda_k \frac{\partial u_k(x)}{\partial c_{k,l}},$$

where $x \in \Re^{(I-1)+m}$. Proposition 3 uses the following regularity condition, which essentially states that the first-order conditions for efficiency in equation (3.5) change smoothly as prices and permit allocations change:

Regularity condition. The matrix of first partial derivatives of the function Ω has full rank.

Note that Ω is defined on a compact set in $\Re^{(I-1)+m}$.

We now distinguish two cases: (1) $(I - 1) + m \leq (I - 1) \times m$ and (2) $(I - 1) + m > (I - 1) \times m$. In case 1 the dimension of the domain of Ω is less than or equal to that of the range, the regularity condition implies that the matrix of first partial derivatives is 1 to 1, and the compactness of the domain implies that the number of zeros of Ω is finite.

In case 2 the dimension of the domain exceeds that of the range. By basic transversality theory, the dimension of a preimage of zero is a manifold of codimension $(I - 1) + m - (I - 1) \times m > 0$ and is therefore a set of measure zero.

Note that an efficient equilibrium will be in the intersection of the graph of $p = p(E)$ with the zeros of Ω. In the case of two regions, there is a simple proof that this intersection is nonempty.

References

1. Brown, D., and G. Heal. "Equity, Efficiency and Increasing Returns." *Review of Economic Studies,* 46, no. 4 (1979): 571–85.
2. Chichilnisky, G. "A Comment on Implementing a Global Abatement Policy: The Role of Transfers." Paper presented at the International Conference on the Economics of Climate Change, OECD/IEA, Paris, June 1993. Published in *The Economics of Climate Change,* ed. Tom Jones (Paris: OECD, 1994), pp. 159–70.
3. Chichilnisky, G., and G. Heal. "Global Environmental Risks." *Journal of Economic Perspectives* 7, no. 4 (fall 1993): 65–86.
4. Chichilnisky, G., and G. Heal. "Who Should Abate Carbon Emissions? An International Perspective." *Economics Letters* 44 (February 1994): 443–49.
5. Chichilnisky, G., and G. Heal. "Efficient Abatement and Marginal Costs." Working paper, Columbia Business School, Columbia University, fall 1993.
6. Chichilnisky, G., and G. Heal. "Implementing the Rio Targets: Perspectives on Market-Based Approaches." Working paper, Columbia Business School, Columbia University, spring 1993.
7. Chichilnisky, G., and G. Heal. "Markets for Tradable CO_2 Emission Quotas: Principles and Practice." Economics Department Working Paper No. 153, OECD, Paris, 1995. Also published as *Topics in Environmental Economics,* ed. M. Boman et al. (Amsterdam: Kluwer Academic Publishers, 1999).
8. Chichilnisky, G., V. Gornitz, G. Heal, D. Rind, and C. Rosenzweig. "Building Linkages among Climate, Impacts and Economics: A New Approach to Integrated Assessment." Working Paper, Goddard Institute for Space Studies, 1997.
9. Coase, R. "The Problem of Social Costs." *Journal of Law and Economics* 3 (1960): 1–44.
10. Coppel, J. "Implementing a Global Abatement Policy: Some Selected Issues." Paper presented at the International Conference on the Economics of Climate Change, OECD/IEA, Paris, June 1993.
11. Dales, J. *Pollution, Property and Prices.* Toronto: University of Toronto Press, 1968.
12. Dasgupta, P., and G. Heal. *Economic Theory and Exhaustible Resources.* Cambridge: Cambridge University Press, 1979.
13. Eyckmans, J., S. Proost, and E. Schokkaert. "Efficiency and Distribution in Greenhouse Negotiations." *Kyklos* 46 (1993) 363–97.

14. Foley, D. "Lindahl's Solution and the Core of an Economy with Public Goods." *Econometrica* 38, no. 1 (1970): 66–72.
15. Grubb, M. *The Greenhouse Effect: Negotiating Targets.* London: Royal Institute of International Affairs, 1989.
16. Hartwick, J. "Decline in Biodiversity and Risk-Adjusted NNP." Working paper, Queen's University, Kingston, Ontario, 1992.
17. Heal, G. "Efficient Abatement and Political Targets." Working paper, Graduate School of Business, Columbia University. Forthcoming in *Proceedings of the Second Nordic Conference on the Environment,* ed. B. Kriström.
18. Heal, G., and Y. Lin. "Efficiency and Equilibrium in Permit Markets." Working paper, Graduate School of Business, Columbia University, fall 1997.
19. Laffont, J.-J. *Effets externes et théorie économique.* Paris: Editions du CNRS, 1977.
20. Lin, Y. "A Two-Country Analysis of Efficient Allocations in Permit Markets." Mimeograph, Department of Economics, Columbia University, fall 1993.
21. Meade, J. "External Economies and Diseconomies in a Competitive Situation." *The Economic Journal* 62 (1952): 54–67. Reprinted in *Readings in Welfare Economics,* ed. Kenneth Arrow and Tibor Scitovsky. London: Allan and Unwin, 1969, pp. 54–67.
22. Milgrom, P., and J. Roberts. *Economics, Organization and Management.* New York: Prentice Hall, 1992, chap. 2.
23. Noll, R. "Implementing Marketable Emission Permits." *American Economic Review Papers and Proceedings* 72 (1982): 120–24.
24. Starrett, D. "Fundamental Nonconvexities in the Theory of Externalities." *Journal of Economic Theory* (1972).
25. Stavins, R. "Transaction Costs and the Performance of Markets for Pollution Control." Faculty Research Working Paper Series, R93-14, John F. Kennedy School of Government, fall 1993.
26. Stavins, R., and R. Hahn. "Trading in Greenhouse Permits: A Critical Examination of Design and Implementation Issues." Faculty Research Working Paper Series, R93-15, John F. Kennedy School of Government, fall 1993.
27. Weyant, J. "Costs of Reducing Global Carbon Emissions: An Overview." *Journal of Economic Perspectives* (fall 1993).

Chapter 4

Emissions Constraints, Emission Permits, and Marginal Abatement Costs

Geoffrey Heal

4.1 Introduction

Should the marginal cost of emission abatement be equalized across countries? Do markets for tradable emission permits lead to Pareto-efficient patterns of emission abatement? Until recently, the standard answers to both questions were yes. However, Chichilnisky [4] and then, in a more general context, Chichilnisky and Heal [5] proved that the efficient abatement of carbon dioxide (CO_2) emissions does not require the equalization of marginal abatement costs across countries. Equalization is required if and only if it is possible to make unrestricted and free lump-sum redistributions of wealth sufficient to equate the marginal social valuation of consumption in all countries. It follows almost immediately that markets for tradable emission permits do not lead in general to Pareto efficiency, as shown in chapter 3. Chichilnisky, Heal, and Starrett's central result there is that if a market for emission rights is introduced, then the manner in which the emission rights are initially distributed between countries is important for efficiency. To be specific they showed that only a finite number of ways of allocating a given total of emission rights between countries will lead to Pareto-efficient outcomes. Distribution and efficiency are linked in

I am grateful to Graciela Chichilnisky, Peter Sturm of the Economics Department of the OECD and Joaquim Oliveira-Martins, Economics Department, OECD for comments and suggestions. Financial support from the OECD, the Global Environment Facility of the World Bank, and NSF grant 93–09610 is also acknowledged. This chapter replaces an earlier paper entitled "Political Targets and Marginal Abatement Costs." This version was written while the author was visiting the Beijer Institute in Stockholm. I am grateful to Karl-Göran Mäler for his hospitality and comments.

competitive economies in which one trades the right to produce privately produced public goods such as carbon dioxide (CO_2) emissions.

Although this point is simple analytically, it has considerable policy implications. For example, prior to this observation it was taken as given that the burden of emission abatement should be borne disproportionately by developing countries by virtue of their supposedly lower marginal abatement costs.[1]

The initial papers (Chichilnisky [4], Chichilnisky and Heal [5], and Chichilnisky, Heal, and Starrett [6]) led to an explosion of interest in these issues. Prat in chapter 6, Heal and Lin in chapter 5, Dwyer [10], Chao and Peck [3], Mäler [12], Mäler and Uzawa [13], Uzawa [16], Manne [14], and Bohm [1] have all subsequently commented on or extended the initial results in various ways. Dwyer, Heal and Lin, and Prat all review issues related to the efficiency of markets for emission permits. Prat looks at the consequences of always distributing permits in a fixed ratio between the participating countries. He shows that, for each set of proportions, there is a total level of emissions such that distributing it in these proportions will lead to Pareto efficiency. Drèze [9] has made a similar observation. Heal and Lin and Dwyer review the implications of strategic behavior in permit markets. The key point here is that in deciding how much to emit in a regime of international emission permits, each country has to make some conjecture about the total levels of emissions produced by all others, as its utility and therefore its demand for permits depends on this. Chichilnisky, Heal, and Starrett (CHS) model a situation in which each country assumes that the total levels of emissions will be that desired by the agency issuing the permits; that is, each country assumes that the international permit regime will be successful in attaining its goals. Heal and Lin and Dwyer look instead at worlds where countries take the emission levels of others as given in the Nash tradition. They show that in these worlds it is more difficult to achieve Pareto efficiency: Heal and Lin show that only a finite number of points on the Pareto frontier can be attained as equilibria with this behavior. Not surprisingly it is easier to attain efficiency if everyone believes that efficiency will be attained and acts accordingly. Chao and Peck and also Manne investigate numerically the interactions between equity and efficiency indicated by the original results of Chichilnisky and Chichilnisky and Heal. Mäler explores the relationship between the CHS results and a Lindahl equilibrium, a more traditional equilibrium concept for market economies with public goods. Many of the counterintuitive results in CHS emerge because a permit market for emissions is a market for a public good but one with uniform prices rather than the individualized prices required in the Lindahl approach. It is therefore an incom-

[1] Much of the rationalization of joint implementation rests on this supposition.

plete market relative to the framework within which Pareto efficiency has been established.

Several commentators have enquired whether equivalent results hold in a framework in which the total level of abatement, instead of being selected as part of an efficient allocation, is imposed arbitrarily by a political authority. Their motivation is a feeling that any global carbon emission targets ultimately selected will reflect political compromise rather than rational economic analysis, so that the relevant policy question is the attainment of efficiency subject to this constraint. We analyze such a situation here. Assuming that an arbitrary level of emission abatement is imposed on the world economy, we ask again the questions with which this chapter opened: Should the marginal cost of emission abatement be equalized across countries? Do markets for tradable emission permits lead to efficient patterns of emission abatement? However, we now ask them in the context of a concept of constrained, or second-best, efficiency.

The answers are exactly as in the previous chapter: Equalization of marginal costs is necessary for constrained efficiency if and only if it is possible to make unrestricted lump-sum transfers of wealth between countries on a scale sufficient to equalize the marginal social valuation of consumption in all countries, and, as a direct consequence, only certain distributions of emission permits are compatible with the attainment of constrained efficiency by way of permit markets. This is an unusual case of first-best results continuing essentially unchanged in a second-best framework.

4.2 The Model

The model and notation are identical to those in Chichilnisky and Heal [5]. The world economy consists of I regions, $I \geq 2$, indexed by $i = 1, ..., I$. Each has a utility function u_i, which depends on its consumption of a vector of m private goods $c_i \in \Re^M$ and on the quality of the world's atmosphere, a, which is a public good. Formally, $u_i(c_i, a)$ measures welfare, where $u_i : R^{M+1} \to R$ is a continuous, strictly concave function and $\partial u_i/\partial c_{i,m} > 0$, $\partial u_i/\partial a > 0$. The quality of the atmosphere, a, is measured by, for example, the reciprocal or the negative of its concentration of CO_2. Let y_i be a vector in R^M giving the production levels of the M private goods in country i. Then the concentration of CO_2 is affected by production:

$$a = \sum_{i=1}^{I} a_i,\ a_i = \Phi_i(y_i) \quad \text{for each country } i = 1, ..., I, \quad \text{and}$$

$$\frac{\partial \Phi_i}{\partial y_{i,m}} < 0 \qquad \forall i \text{ and } m, \qquad (4.1)$$

where a is a measure of atmospheric quality overall and a_i is an index of the abatement carried out by country i. The production functions Φ_i are continuous and show the trade-off between abatement or quality of the atmosphere and the output of consumption goods. An allocation of consumption and abatement across all countries is a vector

$$(c_1, a_1, ..., c_I, a_I) \in \Re^{(M+1)I},$$

as for each of the I regions there are M private goods and one level of abatement. An allocation is feasible if it satisfies constraint (4.1) and the condition that the total consumption of each private good worldwide be equal to the total production, that is,

$$\sum_{i=1,...,I} c_i = \sum_{i=1,...,I} y_i \tag{4.2}$$

Constraint (4.2) allows private goods to be transferred freely between regions; that is, it allows unrestricted lump-sum international redistributions.

An allocation is called feasible with lump-sum transfers if it satisfies the constraints (4.1) and (4.2). It is feasible without lump-sum transfers if it satisfies

$$c_i = y_i \forall i. \tag{4.3}$$

Each region i faces a constraint in terms of allocating total endowments into either consumption c_i or atmospheric quality a_i, represented by the function Φ_i. Then a Pareto-efficient allocation is described by a solution to the problem:

$$\max W(c_1, ..., c_i, a) = \sum_{i=1}^{I} \lambda_i u_i(c_i, a), \tag{4.4}$$

$$a = \sum_{i=1}^{I} a_i, \; a_i = \Phi_i(y_i), \quad \text{for each country } i = 1, ..., I, \quad \text{and}$$

$$\sum_{i=1,...,I} c_i = \sum_{i=1,...,I} y_i. \tag{4.5}$$

Note that the marginal cost of abatement in region i in terms of good m is just the reciprocal of the marginal productivity with respect to m of the function Φ_i:

$$MC_{i,m}(a_i) = -1/\frac{\partial \Phi_i}{\partial y_{i,m}}. \tag{4.6}$$

4.3 Emission-Constrained Efficiency

In this section we introduce the concept of constrained efficiency that we work with here and then establish results about the relationship between equalization of marginal abatement costs and efficiency in this sense.

DEFINITION 1 An allocation is emission-constrained efficient if it maximizes a weighted sum of utilities (4.4) subject to the feasibility constraint either (4.5) (for the case of lump-sum transfers) or (4.3) (for the alternate case) and if in addition it satisfies a constraint on total abatement

$$a = a^* \tag{4.7}$$

specifying a given total abatement level a^*.[2]

4.3.1 Without Lump-Sum Transfers — Chichilnisky [4] and Chichilnisky and Heal [5] established the following proposition concerning first-best Pareto-efficient allocations:

PROPOSITION 1 At a Pareto-efficient allocation $(c_1^*, a_1^*, ..., c_I, a_I^*)$, the marginal cost of abatement in terms of good m in each country, $MC_{i,m}(a_i^*)$, is inversely proportional to the marginal valuation of the private good $c_{i,m}$, $\lambda_i \partial u_i / \partial c_{i,m}$. In particular, the marginal costs will be equal across countries if and only if the marginal valuations of the private good are equal; that is, $\lambda_i \partial u_i / \partial c_{i,m}$ is independent of i.

We now establish a result exactly equivalent to this for the case of emission-constrained efficiency. The only difference in the propositions lies in the replacement of the words "Pareto efficient" by "emission-constrained efficient."

PROPOSITION 2 At an allocation $(c_1^*, a_1^*, ..., c_I^*, a_I^*)$ which is emission-constrained efficient, the marginal cost of abatement in each country in terms of good m, $MC_{i,m}(a_i^*)$, is inversely proportional to the marginal valuation of the private good $c_{i,m}$, $\lambda_i \partial u_i / \partial c_{i,m}$. In particular, the marginal costs will be equal across countries if and only if the marginal valuations of the private good are equal; that is, $\lambda_i \partial u_i / \partial c_{i,m}$ is independent of i.

---

[2]This is formulated as an equality. The results would be unchanged if instead we specified an inequality $a \geq a^*$ with a lower bound for abatement.

PROOF. An emission-constrained efficient allocation, being the solution to the maximization of (4.4) subject to (4.3) and (4.7), must be a stationary point of the Lagrangian

$$L = \sum_k \lambda_k u_k\left(c_k, \sum_k \Phi_k(c_k)\right) + \gamma\left(\sum_k \Phi_k(c_k) - a^*\right),$$

where γ is the shadow price associated with constraint (4.7) on total emissions and so must satisfy the first-order conditions

$$\lambda_i(\partial u_i/\partial c_{i,m}) = -\gamma \frac{\partial \Phi_i}{\partial y_{i,m}} - \frac{\partial \Phi_i}{\partial y_{i,m}} \sum_k \lambda_k(\partial u_k/\partial a) \tag{4.8}$$

for each country $i = 1, ..., I$. Because $MC_{i,m}(a_i^*) = -1/(\partial \Phi_i/\partial y_{i,m})$, the allocation satisfying (4.8) is characterized by

$$MC_{i,m}(a_i^*) = \frac{\gamma + \Sigma_k \lambda_k(\partial u_k/\partial a)}{\lambda_i \partial u_i/\partial c_i}, \tag{4.9}$$

and the proposition follows. ■

Equation (4.9) is identical to the equivalent equation, $MC_i(a_i^*) = \Sigma_k \lambda_k(\partial u_k/\partial a)/(\lambda_i \partial u_i/\partial c_i)$ on page 446 of Chichilnisky and Heal [5] and page 130 of this volume, except for the presence of the term γ reflecting the constraint (4.7). The result is qualitatively the same as in the previous case because, being the shadow price on the provision of a public good, γ is common across all countries.[3] Proposition 2 shows that the product of the marginal valuation of private consumption and the marginal cost of abatement in terms of consumption is equal across countries. Following Chichilnisky and Heal, we write this product as $\lambda_i \partial u_i/\partial c_i \cdot \partial c_i/\partial a$ and note that it can be interpreted as the marginal cost of abatement in country i measured in utility terms, that is, in terms of its contribution to the social maximand $\Sigma_j \lambda_j u_j(c_j, a)$. Equation (4.9) therefore tells us that the marginal cost of abatement in this generalized sense must equal the sum of the marginal valuations of abatement across all countries plus an amount reflecting the shadow price of the abatement constraint. An immediate implication is that in countries that place a high marginal valuation on consumption of the private good (typically low-income countries), the mar-

[3]Of course, if (4.7) is not binding, then $\gamma = 0$, and this condition is precisely the first-order condition characterizing full Pareto efficiency in Chichilnisky and Heal [5]. This will occur only if the specified abatement level a^* is Pareto efficient.

ginal cost of abatement at an efficient allocation will be lower than in other countries. If we assume an increasing marginal cost of abatement (diminishing returns to abatement), then this of course implies lower levels of abatements in poor countries than in rich countries.

Note that these results would be completely unchanged if we were to replace the equality constraint $a = a^*$ by the inequality $a \geq a^*$, placing a lower bound on the acceptable level of abatement. In this case the previous Lagrangean would be unaltered, but the shadow price γ associated with the abatement constraint would satisfy a complementary slackness condition, indicating that it would be zero if the abatement constraint were satisfied with strict inequality. When $\gamma = 0$, we have precisely the results of the previous papers.

4.3.2 With Lump-Sum Transfers — Under what conditions can we recover the conventional wisdom that marginal abatement costs should be equalized across countries? The answer is as in Chichilnisky and Heal [5]: We need to equate the terms $\lambda_i \partial u_i / \partial c_i$ across countries. This could be done by assumption. However, given the enormous discrepancies between the income levels in OECD countries and countries such as India and China and the need for all of them to be involved in an abatement program, such a value judgment seems most unattractive.

There is an alternative possibility. Modify the problem to allow unrestricted transfers of private goods between countries, so that efficiency is defined by maximization of (4.4) subject to the feasibility condition (4.5):

$$\max W(c_1, c_2, c_i, \ldots, a) = \sum_i \lambda_i u_i(c_i, a) \quad \text{subject to}$$

$$a_i = \Phi_i(y_i),\ a = \sum a_i,\ \sum y_i = \sum c_i, \quad \text{and} \quad a = a^*. \tag{4.10}$$

We now require the sum of the consumptions across countries to equal the sum of the productions—$\Sigma\, y_i = \Sigma\, c_i$—instead of having these equal on a country-by-country basis. By this modification we are allowing the transfer of goods between countries; that is, we are allowing lump-sum transfers. Note that this is not a model of international trade, which would require the imposition of balance-of-trade constraints.[4] Clearly, the first-order conditions again require that

$$MC_i = \frac{\Sigma_k \lambda_k (\partial u_k / \partial a) + \gamma}{\lambda_i (\partial u_i / \partial c_i)}, \tag{4.11}$$

[4]See the discussion of this in CHS [6].

but in addition we now require that

$$\lambda_i(\partial u_i/\partial c_i) = \mu \; \forall i. \tag{4.12}$$

Therefore, we now have equalization of marginal abatement costs across countries at the ratio

$$\frac{\Sigma_k \lambda_k(\partial u_k/\partial a) + \gamma}{\mu},$$

where, as before, γ is the shadow price on the total emission (abatement) constraint and μ is that on the constraint equating total output of the private good to consumption. Therfore, if we solve an optimization problem that allows unrestricted transfers between countries and we can and do make the transfers that are needed to solve this problem, it will then be efficient to equate marginal abatement costs, with or without an arbitrary constraint on total abatement. The imposition of an arbitrary constraint on abatement, forcing us into the world of second best, makes no difference to the appropriate relationship between marginal abatement costs. This is because the first-order condition in this case, as in the previous case without lump-sum transfers, differs from that without an arbitrary abatement constraint only in the presence of the shadow price γ in the expression for marginal cost.

4.4 Emission Permits and Emission Constraints

How would the imposition of emission constraints as discussed previously affect the results of the previous chapter on efficiency and the distribution of emission rights? As one might expect, they all carry through again. An immediate implication of the competitive trading of emission permits at a uniform price is the equalization of marginal emission costs, narrowly defined, and if the equality of these marginal costs does not characterize efficiency except for particular distributions of wealth, then the trading of emission permits can be expected to lead to efficiency only for those same particular distributions. Another intuition that leads to the same conclusion was mentioned before: Efficiency in markets for public goods in general requires Lindahl markets with as many prices as there are agents. In the absence of these markets, one cannot expect efficiency, constrained or otherwise.

Formally, let each country be given an allocation E_i of emission rights, where $\Sigma_i E_i = E^*$ and E^* is the agreed total level of emissions worldwide. They can trade these as price takers in a market in which there is a single price

p_e for the right to emit one unit. Countries therefore maximize utility subject to the budget constraint

$$\sum_l c_{i,l} p_l = \sum_l y_{i,l} p_l - p_e(E_i + a_i). \tag{4.13}$$

The interpretation of the right-hand side of this budget constraint is as in the previous chapter: The difference between actual emissions e_i and target emissions E_i is $e_i - E_i = e_i^N - a_i - E_i$, where e_i^N is the emission level of region i when abatement is zero. For simplicity we have dropped the constant terms in e_i^N. This budget constraint requires that for each country the value of consumption equal the value of production plus the net revenue from the sale of permits. Note that (4.13) can be rewritten as

$$\left(\sum_l c_{i,l} p_l - \sum_l y_{i,l} p_l\right) = -p_e(E_i + a_i). \tag{4.14}$$

Here the left-hand side is the difference between the value of domestic consumption and production, that is, the balance of trade. A surplus of consumption over production (i.e., a position of net imports) is funded by the revenue generated by sales of permits in international markets. Conversely, a net purchase of permits in international markets has to be matched by a surplus of production over consumption and therefore a net export position. This interpretation of the budget constraint makes it clear that controlling the initial endowments of emission rights acts as a substitute for lump-sum transfers.

Each country seeks to maximize its utility $u_i(c_i, a)$ subject to the budget constraint (4.13) and to the production relations given in (4.1). We assume that in so doing it supposes the total level of emissions to be fixed at E^*, the desired total level. This in effect implies the existence of a credible intergovernment agency that sets and implements global emission targets. An alternative (explored by Heal and Lin in the next chapter) is to look for a Nash equilibrium in countries' abatement levels.

In the case of a total level of emissions taken by all countries to be E^*, each country chooses consumption levels and abatement or emission levels to satisfy

$$\overbrace{\frac{\dfrac{\partial u_i}{\partial c_{i,l}}}{\dfrac{\partial u_i}{\partial c_{i,m}}} = \frac{p_l}{p_m}}^{\text{mrs} = \text{price ratio}} \tag{4.15}$$

and

$$\overbrace{\frac{\partial \Phi_i}{\partial y_{i,m}} = -\frac{p_m}{p_e}}^{\text{mrt} = \text{price ratio}}. \tag{4.16}$$

These are standard conditions: (4.15) requires only that marginal rates of substitution between goods be equated to their price rations, and (4.16) requires tangency between the production possibility frontier and an isoprofit hyperplane. The latter implies in particular that, for given prices, levels of production (and therefore also of emission) are determined independently of the utility function. (Of course, in equilibrium the prices will depend on preferences.)

How do the first-order conditions (4.15) and (4.16) chosen by the country compare with the conditions (4.11) and (4.12), which describe allocations that are efficient subject to an emission constraint? Clearly, (4.16) is the same as (4.11) provided that

$$\frac{p_m}{p_e} = \frac{\dfrac{\partial u_i}{\partial c_{i,m}}}{\gamma + \Sigma_k \lambda_k \dfrac{\partial u_k}{\partial a}} = \frac{\lambda_k \dfrac{\partial u_k}{\partial c_{k,l}}}{\gamma + \Sigma_k \lambda_k \dfrac{\partial u_k}{\partial a}} \quad \forall k \neq i. \tag{4.17}$$

This condition can only hold if $\partial u_i / \partial c_{i,m}$ and $\lambda_k (\partial u_k / \partial c_{k,l})$ are independent of i and k. Condition (4.12), required for emission-constrained efficiency, automatically implies this. However, there is nothing equivalent in the countries' utility maximization conditions: Condition (4.15) does not imply equalization of marginal valuations.

Therefore, utility maximization subject to the budget constraint (4.13) does not lead to the conditions needed for efficiency. There is an additional requirement represented by (4.12), namely, that $\partial u_i / \partial c_{i,m} = \lambda_k \partial u_k / \partial c_{k,l}$ $\forall l$, $\forall k \neq i$. This condition would of course be satisfied if there were policy instruments available to redistribute resources without restriction across countries—if, for example, lump-sum redistributions were possible. In the absence of such instruments, what is required to ensure that (4.12) is met and that constrained efficiency is attained in the permit market?

Condition (4.12) requires that, for each good, its marginal social valuation be equal for every country. This is a condition with which we are familiar from the previous chapter. As there we note that this is a condition on the distribution of income or wealth. The same arguments as in that chapter can now be applied. We look in more detail at the determinants of the terms $\partial u_i/\partial c_{i,m}$. As $u_i = u_i(c_i, E^*)$, where E^* is fixed, the derivatives of u_i with respect to consumption can depend only on consumption levels. These in turn depend, by means of the budget constraint (4.13), on prices p_l, production levels $y_{i,m}$, abatement levels a_i, and initial endowments of emission rights E_i. Once prices are given, production and abatement levels are fully determined by (4.16). In the absence of policy instruments that can effect unrestricted redistributions across countries, the only variables then available for ensuring that marginal social valuations of consumption are equalized across countries are therefore the initial allocations of permits, and only those initial permit allocations that ensure that (4.12) is satisfied will lead to emission-constrained efficient allocations. We formalize this in the following and show that very few initial allocations satisfy this condition.

PROPOSITION 3 Let E^* be the level of total emissions at an emission-constrained efficient allocation of resources in the economy. Assume that countries maximize utility subject to the budget constraint (4.13) given by the ability to trade emission permits. Assume furthermore that a regularity condition defined below is satisfied. Then, of all possible ways of allocating the total emission E^* among countries as initial endowments, only a subset of measure zero will lead to market equilibria that are emission-constrained efficient. If the inequality $(I-1)+M \leq (I-1) \times M$ holds, then only a finite number of ways of allocating the emission rights lead to efficiency.

REMARK 1 Strict concavity and the regularity assumption are needed for this result. Otherwise, one can construct counterexamples. For example, with quasi-linear preferences of the form $u_i(a) + \alpha_i c_i$, $\alpha_i > 0$, there might be infinitely many allocations of permits that will lead to efficient outcomes.

Consider the first-order conditions for efficiency:

$$\frac{\partial u_i}{\partial c_{i,l}} - \lambda_k \frac{\partial u_k}{\partial c_{k,l}} = 0. \tag{4.18}$$

Define the function Ω from $\Re^{(I-1)+M}$ to $\Re^{(I-1)\times M}$. Its arguments are those of (4.8), namely, E_i, $i = 1, ..., I$ and p_l, $l = 1, ..., M$ and e. Now, as the E_i are nonnegative and sum to a fixed number and there are only M relative prices, Ω is defined on $\Re^{(I-1)+M}$:

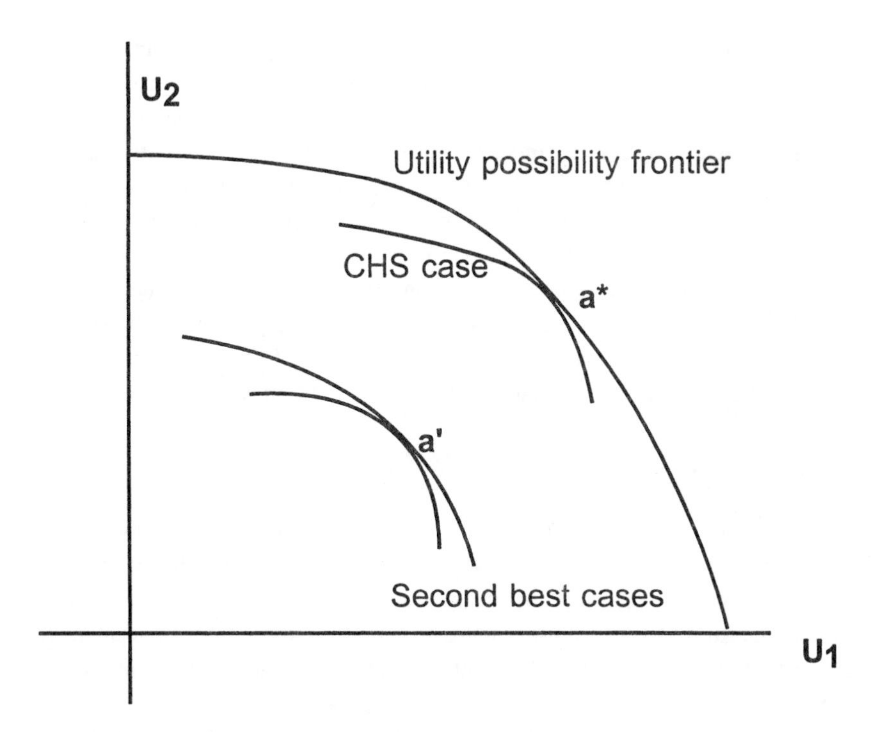

FIGURE 4.1. The frontiers corresponding to different allocations of an arbitrary emis sion total with and without lump-sum transfers.

$$\Omega\colon \Re^{(I-1)+M} \rightarrow \Re^{(I-1)\times M},\ \Omega(x) = \frac{\partial u_i(x)}{\partial c_{i,l}} - \lambda_k \frac{\partial u_k(x)}{\partial c_{k,l}},$$

where $x \in \Re^{(I-1)\times M}$. Proposition 3 uses the following regularity condition, which essentially states that the first-order conditions for efficiency change smoothly as prices and permit allocations change:

> *Regularity condition.* The matrix of first partial derivatives of the function Ω has full rank.

PROOF. The proof copies exactly that in the previous chapter. ■

How does the intuition behind this result relate to the equivalent result in the previous chapter? It can be explained by a very similar diagram (see figure 4.1). The figure repeats figure 3.2 of chapter 3 with additions. It shows the utility possibility frontier of a two-person economy and the utility vectors that emerge from trading permits corresponding to a total level of emissions associated with an efficient equilibrium. The point a′ shows the utility levels at an arbitrary inefficient equilibrium, associated with which is a total level of emissions, say, E_X. The two frontiers through a′ show the utility levels attained by trading emission rights totaling $E_{a'}$ under two different conditions. The outer frontier corresponds to the case in which lump-sum transfers ensure that conditions (4.12) is satisfied, namely, that the marginal valuations of consumption are equal across countries. The inner one corresponds to the case in which this does not happen. Utility vectors on the outer frontier through a′ are constrained Pareto efficient. Those on the other frontier are not, except at the point at which the two frontiers touch. At this point the distribution of emission rights is consistent with constrained efficiency, and no redistribution is needed.

4.5 Conclusions

Efficient abatement subject to an arbitrarily chosen emission level does not in general require equalization of marginal abatement costs; rather it requires equalization of the marginal social opportunity costs of abatement across countries. Marginal costs in the usual sense are to be equalized only if we can make unrestricted lump-sum transfers between countries, not a very interesting hypothesis.

An implication is that for the attainment of emission-constrained efficiency by the trading of emission permits, the initial distribution of permits (property rights) matters, as only a finite number of initial distributions lead to emission-constrained efficiency. The initial allocation of emission permits may play the role of lump-sum transfers: Certain initial distributions of these permits lead to efficiency because they correspond to the lump-sum transfers, which equate marginal valuations of the consumption good, as required for the equalization of marginal costs. The relationship between efficiency and distribution noted in CHS in chapter 3 for the case of Pareto efficiency continues for the case of emission-constrained efficiency.

References

1. Bohm, P. "Should Marginal Carbon Abatement Costs Be Equalized across Countries?" Mimeograph, Department of Economics, University of Stockholm, August 1993.
2. Brown, D., and G. Heal. "Equity Efficiency and Increasing Returns." *Review of Economic Studies* 46, no. 4 (October 1979): 571–85.
3. Chao, H., and S. Peck. "Pareto Optimal Environmental Control and Income Distribution with Global Climate Change." Working paper, Electric Power Research Institute, Palo Alto, California, July 1994.

4. Chichilnisky, G. "The Abatement of Carbon Emissions in Industrial and Developing Countries." Paper presented at the International Conference on the Economics of Climate Change, OECD/IEA, Paris, June 1993. Published in *The Abatement of Carbon Emissions in Industrial and Developing Countries,* ed. Tom Jones. Paris: OECD, 1994.
5. Chichilnisky, G., and G. Heal. "Who Should Abate Carbon Emissions? An International Viewpoint." *Economics Letters* 44 (spring 1994): 443–49.
6. Chichilnisky, G., G. Heal, and D. Starrett. "The Design of Markets for Emission Permits: Equity and Efficiency Are Not Orthogonal." Working paper, Columbia Business School, Columbia University, and Department of Economics, Stanford University, (chapter 3 of this volume).
7. Coppell, J. "Implementing a Global Abatement Policy: Some Selected Issues." Paper presented at the International Conference on the Economics of Climate Change, OECD/IEA, Paris, June 1993. Published in *The Abatement of Carbon Emissions in Industrial and Developing Countries,* ed. Tom Jones. Paris: OECD, 1994.
8. Dasgupta, P., and G. Heal. *Economic Theory and Exhaustible Resources.* Cambridge: Cambridge University Press, 1979.
9. Drèze, J. Private communication, 1994.
10. Dwyer, D. "Who Should Abate Carbon Emissions? A Comment." Working paper, Department of Economics, Columbia University, fall 1993.
11. Heal, G., and Y. Lin. "Equilibrium and Efficiency in International Permit Markets." Working paper, Columbia Business School, Columbia University, 1994.
12. Mäler, K.-G. Unpublished manuscript, Beijer Institute of Ecological Economics, Royal Swedish Academy of Science, Stockholm, 1994.
13. Mäler, K.-G., and H. Uzawa. "Tradable Emission Permits, Pareto Optimality and Lindahl Equilibrium." Discussion paper, Beijer Institute of Ecological Economics, Royal Swedish Academy of Science, Stockholm.
14. Manne, A. "Greenhouse Gas Abatement: Towards Pareto Optimality in Integrated Assessments." Working paper, Department of Operations Research, Stanford University, 1994.
15. Prat, A. "The Efficiency Properties of a Constant-Ratio Mechanism for the Distribution of Tradable Emission Permits." Working paper, Department of Economics, Stanford University, 1993. (Chapter 6 of this volume.)
16. Uzawa, H. "Pareto Optimality, Competitive Equilibrium and Lindahl Equilibrium." Discussion Paper Series No.9, Japan Development Bank Center on Global Warming, 1995.

Chapter 5

Equilibrium and Efficiency: International Emission Permits Markets

Geoffrey Heal
Yun Lin

5.1 Introduction

Climate change poses potential serious problems for our global community. The Intergovernmental Panel on Climate Change (IPCC) [20] predicts that the global mean temperature will rise as much as 3°C above the present value before the end of the twenty-first century if the current trend of greenhouse gas emissions persists.[1] Global warming, experts believe, could cause severe detrimental economic and ecological effects, among them being decreases in agricultural productivity, more frequent storms, and alterations of ecological systems. Although uncertainties[2] still remain in terms of both scientific evidence of the greenhouse effect and the consequences of global warming, the scale, inertia, and possibly irreversible nature of climate change has caught the attention of 157 world leaders, who gathered in Rio de Janeiro in June 1992 to sign the Framework Convention on Climate Change (FCCC), which commits parties to immediate action on the issue.

The greenhouse effect is a typical public goods problem. Gases emitted mix quite uniformly over time in the atmosphere. On the other hand, unlike other

The authors thank Graciela Chichilnisky, Duncan Foley, Bruce Greenwood, and Alex Pfaff for their valuable comments.

[1]Cline [7,8] has a detailed account of the scientific basis of the greenhouse effect.

[2]For a formulation of global environmental problems in the framework of risk analysis, see Chichilnisky and Heal [3].

public goods (e.g., national defense), greenhouse gases (GHGs) are produced privately.[3] Each country's emission contributes a small part of the total emissions; therefore, a unilateral action on the part of an individual country could hardly have an impact on the whole problem. This leads to the concept of coordinated action, or joint implementation (JI), which is a framework that accommodates coordinated national environmental policies within a group of countries to achieve some specified abatement target. Numerous authors have made the point that to cut or stabilize global GHG emissions, a minimal requirement is to bring those major emitters into an agreement.

In a world of consistent disagreement, an agreement is always difficult to reach. Some countries, especially those developing countries, have already voiced their concerns. These countries argue that the accumulation of GHGs in the outer layer of the earth is the result of industrialization of today's developed countries. If any emission cut is to be made, it should be made by those countries, not by the developing countries. The developed countries, on the other hand, agree that they are mainly responsible for the accumulation of those GHGs, but they argue that their effort alone without the cooperation from the South could not solve the problem. Who should abate? Chichilnisky and Heal [1] asked this question. For an answer, we must go back to the principles of economics.

The two standard textbook approaches to public goods problem are taxes/subsidies of the Pigovian tradition and the introduction of property rights of the Coasian tradition.[4] In practice these translate to policies such as emission targets, domestic carbon taxes, international emission taxes, and tradable emission entitlements.[5] Although emission taxes in principle could achieve emission reductions (see Hoel [19]), it is difficult to implement them in practice, especially at the international level. In contrast the Coasian approach has several advantages: The administrative costs are low, markets are easy to organize, and the environmental uncertainty and risks can be decentralized through markets. For these reasons we discuss in this chapter only the Coasian approach to global emission reductions. It is not our claim that environmental tax policies are useless, as taxes could still be an effective option for domestic environmental management.

---

[3] "Knowledge" shares many of the characteristics of the public good "emission abatement" we are discussing here. From a broad perspective knowledge is also produced privately but consumed by all. Thanks to Graciela Chichilnisky for pointing this out. In a separate paper (Lin [22]), one of the authors discusses the implications of knowledge accumulation in an economy with essential resource inputs.

[4] For a review and comparison of these two approaches, see Chichilnisky and Heal [5] and Laffont and Tirole [21].

[5] For detailed discussions of each of these policy instruments, see Hoel [18] and Grubb [12].

The idea of emission trading was first proposed in a book by Dales in 1968 in the context of water pollution control. He argued that by forming a board (water control board) that sets a water quality standard in Ontario and administers a permits trading program, the water quality in the proposed area could be controlled cost effectively. Montgomery [23] provided a rigorous theoretical treatment. Issues related to the design and management of a permit market are examined thoroughly in Tietenberg [25–27], Hahn [13,14], and Hahn and Noll [15,17]. The idea of emission permits trading was not put into practice until quite recently. The United States implemented an emission trading program for sulfur oxides[6] in 1990 after experimenting with a series of semi-cost-effective measures, such as bubbles, netting, offsets, and emission banking.[7] Some of these concepts, offsets, and emission banking, for example, are close to but different from what we call emission permit markets.

Emission permits trading at an international level poses some new questions for economists. The divorce between equity and efficiency is a central feature of classical welfare economics: In the usual competitive model, any distribution of endowments will lead to an efficient allocation of resources.[8] Recently, Chichilnisky, Heal, and Starrett[9] [6] (CHS) showed that this is not true for economies in which the provision of a privately produced public good is controlled by the trading of production quotas. They showed that the way in which a given total number of tradable permits is distributed among polluters will affect the efficiency of the market solution. It had previously been believed that efficiency and distribution are independent in permit markets. This is sometimes referred to as the Coase theorem [9] about the irrelevance of the allocation of property rights to the attainment of efficiency. Using the same CHS model, Prat [24] showed that given a fixed distribution of the initial permits, there can exist only one total emission level that permit markets can obtain.

This chapter refines the findings of CHS. We study in detail the implications for market efficiency of the selection of the initial abatement targets and/or its

---

[6]The trading program is required by the U.S. Clean Air Act Amendments of 1990 (Public Law 101–549), Title IV (Acid Rain Provisions), in an effort to reduce sulfur dioxide emissions by 10 million tons per year, relative to the base year 1980, by the year 2001.

[7]A bubble is a situation in which the owners of the same plant may increase the pollution at one source while making a corresponding reduction at another. Netting, a similar concept to a bubble, is the process by which a remodeled or expanded plant can escape the lengthy reviewing process by the U.S. Environmental Protection Agency (EPA) if the new plant does not increase emissions signicantly. Offsets are sale/purchase of emission quotas between companies. An emissions bank is a credit agency arranging the transaction of emission quotas.

[8]This is not true for second-best situations, in which equity and efficiency are inextricably linked.

[9]Chichilnisky, Heal, and Starrett build on a model introduced in Chichilnisky [1] and developed in Chichilnisky and Heal [2].

initial distribution among countries. In a departure from CHS, we further allow for strategic behavior of abatement participants. This behavioral assumption, as we show here, significantly restricts the role of a policymaking body, even if international income transfers are allowed, in terms of the selection of both an abatement target and the initial allocation of emission permits. In general the Nash equilibrium allocation of resources achieves Pareto efficiency only for a finite number of initial abatement targets and for each target only for a finite number of specific distributions among an infinite number of possibilities.

This chapter is organized as follows. We study the two-good (one private good, one public good) version of the CHS model in sections 5.2 to 5.7. In section 5.2 we give a brief description of our model and derive the Pareto-optimality conditions. After that we study the competitive permits market equilibrium and look for conditions in terms of the distribution of initial emission permits as necessary for market efficiency. We do the same in section 5.4, but in a Nash setting. In section 5.5 we study in more detail the efficient initial permit allocation and countries' equilibrium trade positions. In section 5.6 we extend our results derived in sections 5.4 and 5.5 by allowing for international income transfers. In section 5.7 we briefly comment on permits market efficiency in case of individual permits pricing (Lindahl equilibrium). Extension to M private goods is done in section 5.8. Most of our results survive as more private goods are added. Section 5.9 concludes.

5.2 Pareto-Efficient Allocation of Resources

Our framework originated in Chichilnisky [1] and was developed in Chichilnisky and Heal [4]. A fuller version is in CHS. For illustration we work first with the one-private-good version of the model and then, in section 5.9, extend the model to M private goods. The world economy consists of I ($I \geq 2$) countries, each endowed with one private consumption good $\bar{Y}_i$ and an abatement technology

$$a_i = \Phi_i(\bar{Y}_i - y_i), \tag{5.1}$$

which transforms the private good into abatement. The term a_i in (5.1) is country i's abatement of carbon dioxide (CO_2) emissions, and y_i is the private good available for consumption. We assume that $\Phi_i(0) = 0$, so all initial private endowment $\bar{Y}_i$ is available for consumption; that is, $y_i = \bar{Y}_i$ if there is no abatement. Some of this private good $\bar{Y}_i$ may be given up to provide a better atmosphere. In addition each country has a utility function $u_i(c_i, a)$, where $a =$

$\Sigma_i a_i$ is the total abatement and c_i is the private goods actually consumed. Finally, we assume that both $u_i(c_i, a)$ and $\Phi_i(.)$ are twice continuously differentiable, strictly concave, and increasing.

DEFINITION 1 Let $\pi_i = a_i/a$ be the actual share of emission abatement contributed by each country i. An allocation $(\{\hat{c}_i\}_{i=1,...,I}, \{\hat{y}_i\}_{i=1,...,I}, \{\hat{\pi}_i\}_{i=1,...,I}, \hat{a})$ is Pareto efficient if it is feasible and cannot be improved on weakly for all countries and strictly for some country i.

Feasibility in our case requires

$$\sum_{i=1}^{I} \hat{\pi}_i = 1 \tag{5.2}$$

and

$$\sum_{i=1}^{I} \hat{y}_i = \sum_{i=1}^{I} \hat{c}_i. \tag{5.3}$$

Note that the feasibility constraint (5.3) here permits unrestricted transfers of consumption between countries. Pareto-efficient allocations are the set of all solutions that maximizes the weighted welfare

$$W(c_1, ..., c_I, a) = \sum_{i=1}^{I} \lambda_i u_i(c_i, a) \tag{5.4}$$

of I countries subject to equations (5.2) and (5.3). As usual we consider all possible welfare weights $\lambda_i \geq 0$ subject to $\Sigma_{i=1}^{I} \lambda_i = 1$. Nonnegativity constraints are understood wherever appropriate.

The following notations will simplify our presentation:

$$MRS_i(a, c_i) = \frac{\partial u_i/\partial a}{\partial u_i/\partial c_i},$$

$$MC_i(a_i) = -\frac{1}{\partial \Phi_i/\partial y_i}.$$

The second line simply says that the marginal cost, in terms of the private good, of producing a_i is just the inverse of the marginal productivity of that good at the abatement level a_i.

Finally, we need a formula to transform the measure of emission abatement into permits. For each country the two measures are clearly negatively related.

We write

$$e_i = e_i^0 - a_i, \qquad (5.5)$$

where e_i^0 (a constant) is interpreted as the natural emission level, or the level at which no preventive measures are taken for pollution control, $y_i = \bar{Y}_i$. With expression (5.5), we therefore may use the terms *emissions* and *abatement level* interchangeably.

LEMMA 1 The marginal cost of emission abatement for each country at an efficient allocation equals the sum of the marginal rates of substitution (between emission abatement and the private good) of all I countries [1,4].

PROOF. The lemma implies that marginal costs are equalized across countries at a Pareto-efficient allocation. This follows from the possibility of unrestricted transfers between countries (see Chichilnisky and Heal [4] and chapter 7). Lemma 1 is derived directly from the first-order conditions to the maximization of problem 5.4. It is a simple exercise to verify that

$$MC_i(a_i) = MC_j(a_j), \qquad \forall\, i \neq j \qquad (5.6)$$

and

$$\sum_{i=1}^{I} MRS_i(a, c_i) = MC_j(a_j) \quad \text{for any } j \qquad (5.7)$$

are necessary conditions for Pareto optimality. ■

Equation (5.7) is the well-known Lindahl-Bowen-Samuelson condition in the public goods literature. It is really another version of marginal benefits equaling marginal cost. It says that the cost of producing one more unit of public good a by one country (it does not matter which one as long as [5.6] is satisfied) must equal the summation of marginal benefits received by all who consume it. For obvious reasons, equation (5.6) is sometimes called the production efficiency condition, whereas equation (5.7) is called the allocation efficiency condition. A Pareto-efficient allocation has the following property.

LEMMA 2 Let $(\{\hat{c}_i\}_{i=1,\dots,I}, \{\hat{y}_i\}_{i=1,\dots,I}, \{\hat{\pi}_i\}_{i=1,\dots,I}, \hat{a})$ be a Pareto-efficient allocation. If there exists another allocation $(\{c'_i\}_{i=1,\dots,I}, \{y'_i\}_{i=1,\dots,I}, \{\pi'_i\}_{i=1,\dots,I}, \hat{a})$ that is also Pareto efficient at the same total abatement level $\hat{a}$, then $\hat{\pi} = \pi'_i$, $\hat{y}_i = y'_i$ for all i.

PROOF. First note that if $\hat{\pi} = \pi'_i$ for all i, then $\hat{y}_i = y'_i$ for all i because Φ_i maps from $\Re$ to $\Re$ and Φ_i is monotonic. We prove the lemma by contradiction. Suppose not. Because $\Sigma_{i=1}^{I} \hat{\pi}_i = \Sigma_{i=1}^{I} \pi'_i = 1$, there must exist at least one j and one k such that $\hat{\pi}_j > \pi'_j$ and $\hat{\pi}_k < \pi'_k$, or equivalently, $a_j^* > a'_j$ and $\hat{a}_k < a'_k$. Then, by concavity of functions $\Phi_i(.)$, we have

$$MC_j(\hat{a}_j) > MC_j(a'_j) \quad \text{and} \quad MC_k(\hat{a}_k) < MC_k(a'_k),$$

which contradicts the necessary condition for Pareto efficiency, $MC_j(a'_j) = MC_k(a'_k)$. This lemma establishes a one-to-one relationship between the total abatement $\hat{a}$ and the actual shares of the total abatement by individual countries $\{\hat{\pi}_i\}$. ■

The lemma further implies that the relationship between $\hat{a}$ and private production levels $\{\hat{y}_i\}$ at the Pareto-efficient allocation is also uniquely determined. Therefore, a Pareto-efficient abatement level uniquely determines the production side of the economy. Note that this result relies only on the production efficiency condition. It has nothing to do with the allocation efficiency condition, which we discuss later. Also note that if there is more than one private good, lemma 2 becomes invalid: neither $\{\hat{\pi}_i\}$ nor $\{\hat{y}_i\}$ can be uniquely determined from $\hat{a}$. This complication is discussed in section 5.8.

In the next two sections, we study emission permit markets. We first consider competitive markets in which countries take the initial abatement target as given.

5.3 Competitive Emission Permits Market

The concept of international emission permits trading is simple. A global emission level, $\overline{e}$, is chosen, and a total of $\overline{e}$ permits is issued and distributed (according to some agreed formula) among the I participating countries. A country holding an initial permit allocation $\overline{e}_i$ (allowances) and emitting e_i may sell the excess $(\overline{e}_i - e_i)$, if $\overline{e}_i$ exceeds e_i, at the price p_e in exchange for private consumption goods, or may buy $(e_i - \overline{e}_i)$ permits by selling consumption goods should the country need more. The price of the private good is taken to be one.

Country i faces a trade-off between emission and private consumption,

$$c_i - y_i = (\overline{e}_i - e_i)p_e$$

or, equivalently, in terms of abatement by using the measure transformation (5.5),

$$c_i - y_i = (a_i - \bar{a}_i)p_e. \tag{5.8}$$

Each country maximizes its utility

$$\max\ u_i(c_i, a) \tag{5.9}$$

subject to budget constraint (5.8) and production feasibility constraint $a_i = \Phi_i(\bar{Y}_i - y_i)$. Each country assumes that the aggregate emissions target is met. This is a key part of the definition of a competitive equilibrium. At the equilibrium we need, of course, equality of the aggregate production and consumption of private goods, that is,

$$\sum_{i=1}^{I} c_i = \sum_{i=1}^{I} y_i. \tag{5.10}$$

Walras's law ensures that the permits market clears as well.

Let $\theta_i = \bar{a}_i/\bar{a}$ be the initial share of abatement assigned to country i. We can rewrite the budget constraint (5.8) as

$$c_i - y_i = (\pi_i\bar{a} - \theta_i\bar{a})p_e = (\pi_i - \theta_i)\bar{a}p_e. \tag{5.11}$$

Notice embedded in this equation the private-good market clearance condition. This is evident, as $\Sigma_i(c_i - y_i) = [\Sigma_i(\pi_i - \theta_i)]\bar{a}p_e = 0$.

DEFINITION 2 Given an abatement target $\bar{a}$ and an initial distribution of the abatement, $\{\theta_i\}_{i=1,\dots,I}$, $0 \le \theta_i \le 1$, $\Sigma\,\theta_i = 1$. An allocation and a price $(\{c_i^*\}_{i=1,\dots,I}, \{y_i^*\}_{i=1,\dots,I}, \{\pi_i^*\}_{i=1,\dots,I}, p_e^*)$ is a competitive equilibrium if for each country i $(c_i^*, y_i^*, \pi_i^*, p_i^*)$ solves problem (5.9).

If a is taken by all countries as constant, as is in the case of CHS, then the first-order conditions to problem (5.9) are simply

$$MC_i(a_i) = p_e, \qquad \forall\, i, \tag{5.12}$$

a well-known condition.

The following lemma about the existence of a competitive equilibrium is easily established.[10]

[10] A proof to the lemma is also shown in Prat [24].

LEMMA 3 Given $\bar{a}$ and $\{\theta_i\}_{i=1,\ldots,I}$, there exists a unique competitive equilibrium.

PROOF. We show the existence and uniqueness by actually solving for the equilibrium. Because $\bar{a}$ is fixed and from (5.12) $MC_i(a_i) = p_e$ for $\forall i$, the actual abatement shares $\{\pi_i^*(\bar{a})\}_{i=1,\ldots,I}$ can be uniquely determined according to lemma 2. Price $p_e^*(\bar{a})$ then is also determined by (5.12) because $a_i^* = \pi_i^*(\bar{a}) \cdot \bar{a}$ is known. From $a_i = \Phi_i(\bar{Y}_i - y_i)$, $y_i^*(\bar{a})$ in turn is calculated. Finally, from (5.11) we have $c_i^* = y_i^* + (\pi_i^* - \theta_i)\bar{a}p_e^*$. Equilibrium $[c_i^*(\{\theta_i\}, \bar{a}), y_i^*(\bar{a}), \pi_i^*(\bar{a}), p_e^*(\bar{a})]$ thus solved is unique. ■

The next two propositions reveal how the selection of an abatement target and its initial distribution affect the efficiency property of a competitive equilibrium. In proposition 1, $\bar{a}$ is assumed to be fixed and θ_i is allowed to change. In proposition 2, the opposite is true.

PROPOSITION 1 Assume that $\bar{a}$ is given. A distribution of initial abatement $\{\theta_i\}_{i=1,\ldots,I}$ associated with $\bar{a}$ leads to a Pareto-efficient allocation of resources if and only if the equilibrium $[c_i^*(\{\theta_i\}, \bar{a}), y_i^*(\bar{a}), \pi_i^*(\bar{a}), p_e^*(\bar{a})]$ satisfies condition (5.7), or

$$\sum_{i=1}^{I} MRS_i(\bar{a},\ c_i^*) = MC_j(a_j^*) \quad \text{for some } j.$$

PROOF. A permit market equilibrium allocation is Pareto efficient if both conditions (5.6) and (5.7) are met at the equilibrium. The marginal cost equalization condition (5.6) clearly is implied by the permits market equilibrium condition (5.12). However, the allocation efficiency condition (5.7) is not guaranteed. To achieve efficiency the equilibrium associated with an initial allocation rule $\{\theta_i\}$ must satisfy (5.7). ■

In the absence of a public good the distribution of initial property rights does not matter in terms of efficiency. Any distribution $\{\theta_i\}$, where $\Sigma_{i=1}^I \theta_i = 1$, would lead to a Pareto-efficient allocation of resources. This is the essence of the Coase theorem. What proposition 1 says is that this is not true if a public good is privately produced. In fact an extra condition (5.7), other than the physical constraint $\Sigma_{i=1}^I \theta_i = 1$, must be imposed on the distribution $\{\theta_i\}$. The selection space for θ_i is reduced by one dimension because of this constraint.

COROLLARY 1 Assume that $\bar{a}$ is given. If there are only two countries, then generically only one distribution, $\{\theta_1, \theta_2\}$, of the initial abatement target $\bar{a}$ leads to a Pareto-efficient allocation of resources.

PROOF. Under the two-country assumption, two equations, constraint (5.7) and the condition $\theta_1 + \theta_2 = 1$, generally would lead to a unique solution of $\{\theta_1, \theta_2\}$. ■

We further claim that if the two countries have the same production and utility functions, then equation (5.7) must be identical to $\theta_1 + \theta_2 = 1$. In this case any distribution of initial permits will lead to Pareto efficiency.

PROPOSITION 2 Assume that $\{\theta_i\}_{i=1,\ldots,I}$ is given. There exists a unique $\overline{a}$, such that the competitive equilibrium $[c_i^*(\{\theta_i\}, \overline{a}), y_i^*(\overline{a}), \pi_i^*(\overline{a}), p_e^*(\overline{a})]$ associated with initial $\overline{a}$ and $\{\theta_i\}_{i=1,\ldots,I}$ is Pareto efficient [24].

PROOF. A detailed proof is given in Prat [24]. Here we give a sketch of the proof.

As stated in proposition 1, a permits market equilibrium achieves Pareto efficiency if

$$\sum_{i=1}^{I} MRS_i[c_i^*(\{\theta_i\}, \overline{a}), \overline{a}] = MC_j(a_j^*) \quad \text{for some } j.$$

This equation contains only one unknown, $\overline{a}$. The proof is complete if we show that the left-hand side of the equation is a decreasing function of $\overline{a}$, as we already know $MC_j(a_j^*)$ increases with $\overline{a}$. ■

5.4 Nash Equilibrium

In this section we depart from CHS by assuming that each country maximizes its utility, taking the abatements of all other countries as given. This means that in solving (5.9), each country i takes a_j ($\forall j \neq i$) as given, or

$$\max\ u_i(c_i, a_i + \Sigma_{j\neq i}\, a_j) \quad \text{subject to} \quad c_i - y_i = (a_i - \overline{a}_i)p_e.$$

The necessary condition for this is

$$MC_i(a_i) = p_e + MRS_i, \qquad \forall\, i, \tag{5.13}$$

which means that, at the equilibrium, the market price for emission permits will always be higher than even the lowest marginal cost of abatement. It is also clear from equation (5.13) that marginal cost equalization—the condition required for Pareto efficiency—is not required for a Nash equilibrium. Therefore, the equilibrium allocation of resources might not be Pareto efficient. Naturally, we would ask, Could we select an abatement target and a distribution formula for the number of permits corresponding to this target such that the equilibrium resource allocation coincides with a Pareto-efficient allocation? This question indeed points directly to the most sensitive issue of any potential international abatement agreement: equity. It also suggests that we could not separate equity from efficiency.

We answer this question by taking the following approach. Pick an arbitrary total abatement level $\bar{a}$ and distribute it arbitrarily among all the member countries. Solve problem (5.9) for equilibrium abatement levels. Check to see whether the equilibrium resource allocation is Pareto efficient. Repeat the process. We are interested in the set of all possible distributions of all possible totals of emission permits that lead to Pareto-efficient outcomes. It turns out that such a set contains only one point: For only one specific total abatement level and one way of distributing it among the I countries could the permits market lead to efficiency.

Pareto efficiency requires both production and allocation to be efficient. Marginal cost equalization across countries (condition [5.6]) ensures production efficiency, and marginal benefits equaling marginal cost (condition [5.7]) ensures allocation efficiency.

We now turn to the allocation efficiency condition (equation [5.7]). As will be shown in the following, for a permit market Nash equilibrium to be compatible with this condition, we must have $MRS_i = MRS_j$, which refines the set of efficient abatement allocations and reduces the choices of aggregate abatement levels. The next lemma shows that the refined set contains at most one point.

LEMMA 4 Among all possible Pareto-efficient allocations, only one allocation $(\{c_i^*\}_{i=1,\ldots,I}, \{y_i^*\}_{i=1,\ldots,I}, \{\pi_i^*\}_{i=1,\ldots,I}, a^*)$ satisfies the condition $MRS_i = MRS_j, \forall i, j$.

PROOF. Consider a Pareto-efficient allocation at which $\forall\, i, j, MRS_i = MRS_j$.

By efficiency and lemma 1, $\Sigma_i MRS_i = MC_i \,\forall\, i$. Now consider an alternative efficient allocation at which the aggregate emission level is greater. Assume contrary to the lemma that once again $\forall\, i, j, MRS_i = MRS_j$. For all countries the abatement level will be greater (because marginal costs are equal, so that all abatement levels move together) and the production of the consumption good lower, and therefore by the concavity assumptions the marginal costs MC_i will be greater for all i. By the assumption that $\forall\, i, j, MRS_i = MRS_j$, the greater marginal costs imply that $\forall i, MRS_i$ is greater. However, for each country abatement is greater and in aggregate consumption lower. This implies that for at least one country i, abatement has risen and consumption has fallen, so that MRS_i has fallen, a contradiction. ■

PROPOSITION 3 Only at a unique total abatement level, and with a unique way of distributing it among the countries as their initial endowments, could the permits market equilibrium lead to Pareto efficiency.

PROOF. Let $(\{c_i^*\}_{i=1,\dots,I}, \{y_i^*\}_{i=1,\dots,I}, \{\pi_i^*\}_{i=1,\dots,I}, a^*)$ be the unique resource allocation under the condition of lemma 4. We prove proposition 3 by construction in two steps.

STEP 1 We show that if a permits market equilibrium exists and its allocation is Pareto efficient, then the resource allocation at the equilibrium must be $(\{c_i^*\}_{i=1,\dots,I}, \{y_i^*\}_{i=1,\dots,I}, \{\pi_i^*\}_{i=1,\dots,I}, a^*)$. Recall the Pareto-efficiency condition

$$MC_i(a_i) = MC_j(a_j), \qquad \forall\, i, j$$

and the permits market equilibrium condition

$$MRS_i(c_i, a) = MC_i(a_i) - p_e, \qquad \forall\, i.$$

An equilibrium, if it exists and is Pareto efficient, must meet both of these conditions, which would require that

$$MRS_i = MRS_j, \qquad \forall\, i, j.$$

However, this is exactly the condition required by lemma 4 leading to the unique resource allocation $(\{c_i^*\}_{i=1,\dots,I}, \{y_i^*\}_{i=1,\dots,I}, \{\pi_i^*\}_{i=1,\dots,I}, a^*)$.

STEP 2 We construct, using the permit markets equilibrium conditions, a unique price and a unique allocation of initial permits. From equation (5.13) and the budget balance condition (5.11), we easily have

$$p_e^* = MC_i(\pi_i^* \mathrm{a}^*) - MRS_i(c_i^*, a^*),$$

$$\theta_i^* = \pi_i^* - \frac{(c_i^* - y_i^*)}{p_e^* a^*}. \qquad (5.14)$$

Therefore, we conclude $(\{c_i^*\}_{i=1,\ldots,I}, \{y_i^*\}_{i=1,\ldots,I}, \{\pi_i^*\}_{i=1,\ldots,I}, p_e^*)$ as the only equilibrium candidate and indeed the only Pareto-efficient equilibrium if the nonnegativeness restriction on the initial permits allocation is relaxed. This completes the proof for proposition 3. ■

A negative initial assignment of abatement target, θ_i^*, to country i would mean a credit to that country in the sense that the country has not only no obligation for its share of abatement but also the rights to claim for cash income with the mere participation in the agreement.

5.5 Characterization of Efficient Permit Markets

What we have developed so far is purely an efficiency argument. It turned out, surprisingly, that there is not much choice—no choice to be precise—for a social planner in terms of choosing either the social optimal emission target or the formula for distributing the initial permits after a target has been chosen. Both the social optimal emission level and the formula for initial permits distribution are uniquely determined following the procedures outlined in the previous section. The rest is left to the markets to decide. Under the behavior assumptions we have made about the agents, the markets should come to an efficient outcome.

We now look into the efficient permit markets arrangement in more detail. As we have said earlier, equity issue is crucial to the success of the permit markets proposed so far. More specifically, each participant of the abatement agreement will be interested in knowing who gets what share of the total abatement assignment. We postpone the direct answer to this question. Instead we ask, What will be the market position of each member of the abatement agreement at the efficient equilibrium outcome? Who are the buyers? The sellers? First we need a few definitions.

DEFINITION 3 Suppose that i and j have the same initial private endowment. Country i is said to have a more *efficient* abatement technology than j if $MC_i(y_i) < MC_j(y_i), \forall\, y_i$.

DEFINITION 4 Let $\{c_i^*\}_{i=1,\ldots,I}$ be the Pareto-efficient consumption allocation at a^* such that $MRS_i = MRS_j$. Country i is said to be more *environment conscious* than j if $c_i^* > c_j^*$.

DEFINITION 5 The unique total abatement level (emissions) as stated in proposition 3 is called *efficient abatement level (emissions);* the unique initial distribution of the efficient abatement (emissions) as stated in proposition 3 is called *efficient initial abatement distribution (permits distribution).*

Definition 3 says that an abatement technology of one country is better than another if the marginal cost of abatement is smaller for every level of private production. This is clearly a global definition. Also notice that this definition is based on the assumption that the two countries have the same initial endowment $\bar{Y}$. To compare two abatement technologies with different initial private endowments, we break the difference into two components. One is the efficiency component, as is indicated in definition 3; the other is the income component, or income effect. A change in the initial private endowment would shift the trade-off curve along the y-axis in the *a-y* space. Marginal cost at a given abatement level for given abatement technology will not change as the initial private endowment changes. Clearly, this definition of abatement technology efficiency does not rank all possible abatement technologies. Our results presented therefore will be indicative rather than comprehensive.

Definition 4, in contrast, is a pointwise one. It says simply the country that values more the same unit of the public good is the one that cares more about the environment. It is recognized in the definition that the consumption allocation at the *efficient abatement level* a^* is unique under the assumption of equalization of marginal rates of substitution (lemma 4).

The last definition is self-evident.

PROPOSITION 4 At an equilibrium with an *efficient initial permits distribution,*

1. of two countries with the same initial endowment in private goods and the same abatement technology, the country that is more *environment conscious* is a relative permits seller;
2. of two countries with the same abatement technology and the same *environment consciousness,* the country with less initial private endowment is a relative permits seller; and
3. of two countries with the same initial endowment in private goods and the same *environment consciousness,* the country that is more *efficient* in its abatement technology is a relative permits seller.

PROOF. Substitute $e_i = e_i^0 - a_i$ into equation (5.14) and rewrite it as

$$\bar{e}_i^* - e_i^* = -(\theta_i^* - \pi_i^*)a^* = \frac{(c_i^* - y_i^*)}{p_e^*}.$$

Country i exports emission permits if the right-hand side of the equation is positive. Country i, in a relative sense, will export more permits than j if

$$(\bar{e}_i^* - e_i^*) - (\bar{e}_j^* - e_j^*) = \frac{(c_i^* - y_i^*) - (c_j^* - y_j^*)}{p_e^*} > 0.$$

Proposition 4 is a restatement of this equation in terms of each country's relative initial private endowment, environment consciousness, and efficiency in abatement technology.

We make one observation before checking through each item of the proposition. If the abatement technology of country i is more efficient than country j, assuming that they have the same initial private endowment, then at the efficient permits market equilibrium, $y_i^* < y_j^*$. A simple argument shows why. Remember that an efficient permits market equilibrium requires that $MC_i = MC_j$. The abatement technology efficiency definition says that at every y, $MC_i < MC_j$. Because marginal cost is a decreasing function of y, $MC_i = MC_j$ holds only if $y_i^* < y_j^*$.

1. Because the two countries have the same endowment and abatement technology, their actual abatement level and therefore outputs y_i^* at the permits market equilibrium must be the same. By referring to the last equation, we know that the actual consumption level of each country will determine who, in a relative sense, exports more permits. The country that is more environment conscious, a higher c^*, will have more excess of permits.

2. Again the same environment consciousness implies that the two countries have the same consumption, $c_i^* = c_j^*$. The same abatement technology would confirm that the actual abatement levels of the two countries are also the same. Therefore, the country that has less initial private endowment will be the one that has less output of private good y^* at the equilibrium, and this country will export more permits.

3. The same environment consciousness implies that the two countries have the same consumption, $c_i^* = c_j^*$. Because the initial endowments of private goods are also the same, the observation at the beginning of the proof says that the country that is equipped with a better abatement technology has a lower output y^* and therefore more excess of permits.

This completes the proof of proposition 4. ■

Next we move to answer the question posed at the beginning of this section: Who is to get what share of the total initial permits? Or, equivalently, who is to be assigned to what share of the total initial abatement?

The difference in abatement assignment two countries *i* and *j* would receive, according to equation (5.14), is

$$\theta_i^* - \theta_j^* = (\pi_i^* - \pi_j^*) - \frac{(c_i^* - y_i^*) - (c_j^* - y_j^*)}{p_e^* a^*}. \qquad (5.15)$$

The next proposition summarizes the distribution of initial abatement assignment as is determined by each country's abatement technology, environment consciousness, and initial endowment of private goods.

PROPOSITION 5 At the market equilibrium with an *efficient initial permits distribution,*

1. of two countries with the same initial endowment in private goods and the same abatement technology, the country that is more *environment conscious* will receive less abatement assignment;
2. of two countries with the same abatement technology and the same *environment consciousness,* the country with less initial private endowment will receive less abatement assignment; and
3. of two countries with the same initial endowment in private goods and the same *environment consciousness,* the country that is less *efficient* in its abatement technology will receive less abatement assignment, assuming that the number of participating members in the abatement agreement is sufficiently large.

PROOF.

1. Because the two countries have the same abatement technology, by definition 1, $\pi_i^* = \pi_j^*$. Further, we have $y_i^* = y_j^*$ because of the equality of their endowments $\bar{Y}_i$ and $\bar{Y}_j$. Equation (5.15) is then reduced to $\theta_i^* - \theta_j^* = -[(c_i^* - c_j^*)/p_e^* a^*]$, which means that the country that is more environment conscious (i.e., higher c_i^*) will be assigned less abatement.
2. The same environment consciousness and same abatement technology, under our definitions, means that $c_i^* = c_j^*$ and $a_i^* = a_j^*$. Further, $\bar{Y}_i < \bar{Y}_j$ implies that $y_i^* < y_j^*$ because $MC_i = MC_j$. It is then evident that the country that has less initial private endowment will have less private good *y* and receive less abatement assignment.
3. The same environment consciousness, by definition 2, implies that $c_i^* = c_j^*$. The difference in initial abatement assignments is then reduced to

$$\bar{a}_i^* - \bar{a}_j^* = (a_i^* - a_j^*) + \frac{(y_i^* - y_j^*)}{p_e^*} = \frac{(p_e^* a_i^* + y_i^*) - (p_e^* a_j^* + y_j^*)}{p_e^*}.$$

Substituting p_e^* by $(I - 1/I)\,MC^*$ and rewriting the last part of the previous equality, we have

$$\bar{a}_i^* - \bar{a}_j^* = \frac{\frac{I-1}{I} MC^*(a_i^* - a_j^*) - (y_j^* - y_i^*)}{p_e^*}.$$

Let i be the country with a more efficient abatement technology, and let $\hat{a}_i = \Phi_i(Y - y_j^*)$. By definition 1, $\hat{a}_i > a_j^*$. We need to prove that $\bar{a}_i^* - \bar{a}_j^* > 0$, which will be true when I is sufficiently large if we can show that $MC^*(a_i^* - a_j^*) - (y_j^* - y_i^*) > 0$ or that $(MC^* a_i^* + y_i^*) - (MC^* a_j^* + y_j^*) > 0$. The last inequality is indeed true because

$$MC^* a_i^* + y_i^* > MC^* \hat{a}_i^* + y_j^* > MC^* a_j^* + y_j^*.$$

This completes the proof of proposition 5. ■

Proposition 5 provides some insights about efficient initial abatement assignments. A preliminary judgment seems to suggest that the allocation formula of initial abatement will not be biased against either developed or developing countries. A typical less developed country is characterized by less environment consciousness and a less efficient abatement technology, two factors that cancel each other out in terms of their roles in determining the initial abatement assignments. The size of initial endowment of private goods, an attribute that could go to either developed countries or developing countries, although more likely to the latter, is positively correlated to the initial abatement assignment.

Take the example of China and the United States. The former would receive more abatement assignment on the basis of its *environment consciousness.* On the other hand, because China has less initial private endowment and is less *efficient* in its abatement technology, our proposition would suggest that more initial abatement be assigned to China. The net result is not clear without real numbers plugged in to the efficient allocation formula.

5.6 Emission Permits and/or Income Transfers

So far we have limited the policy instrument to the distribution of initial permits only. We add one more instrument to the toolbox of a policymaker, who is now not only in charge of distributing the initial permits but also allowed to shift initial private endowments. We consider the efficiency of permits market under the new expanded policy space. In particular we would like to know how the added authority to the policymaker would change (if at all) proposition 3 of section 5.5.

Mathematically, a social planner is to choose an abatement target $\tilde{a}$, its initial distribution $\{\tilde{a}_i\}$, and a reallocation of initial private endowments $\{\tilde{Y}_i\}$ such that

$$\sum_i \tilde{a}_i = \tilde{a} \quad \text{and} \quad \sum_i \tilde{Y}_i = \bar{Y}$$

and each country maximizes its own utility

$$\max u_i(c_i, a) \quad \text{subject to} \quad a_i = \Phi_i(\tilde{Y}_i - y_i), c_i - y_i = (a_i - \tilde{a}_i)p_e, \tag{5.16}$$

the same setting as in section 5.5. The necessary conditions for the maximization of utilities, of course, are also the same as before:

$$MC_i = p_e + MRS_i, \qquad \forall\, i.$$

The new permits market equilibrium $(\{\hat{c}_i\}_{i=1,\dots,I}, \{\hat{y}_i\}_{i=1,\dots,I}, \{\hat{\pi}\}_{i=1,\dots,I}, \hat{p}_e)$ solves problem (5.16) for each country and meets the market-clearing condition $\Sigma_{i=1}^{I} c_i = \Sigma_{i=1}^{I} y_i$.

Let $(\{c_i^*\}_{i=1,\dots,I}, \{y_i^*\}_{i=1,\dots,I}, \{\pi_i^*\}_{i=1,\dots,I}, p_e^*)$ denote the unique Nash equilibrium associated with the fixed initial income distribution $\{\bar{Y}\}$ and the efficient initial abatement distribution $\{\theta_i^*\}$ of a^*. This is the equilibrium we discussed in the previous two sections. The next proposition establishes the connection between this equilibrium and the equilibrium $(\{\hat{c}_i\}_{i=1,\dots,I}, \{\hat{y}_i\}_{i=1,\dots,I}, \{\hat{\pi}\}_{i=1,\dots,I}, \hat{p}_e)$ under the new expanded policy space.

PROPOSITION 6 Even if an authority has control over both the initial permits distribution and the private endowment reallocation, it is still the case that a permits market equilibrium will lead to an efficient allocation if and only if $\tilde{a} = a^*$, where a^* is the efficient abatement level. Furthermore, abatement levels and the reallocation of initial private endowments must satisfy $(\hat{\theta}_i - \theta_i^*) = (\tilde{Y}_i - \bar{Y}_i)/p_e^* a^*$.

PROOF. The proof is basically the same as the one to proposition 3. The only thing that is crucial to the proof of this proposition is the fact that the Pareto-efficient conditions (5.6) and (5.7) are independent of the initial private-goods distribution $\{\bar{Y}_i\}$. We provide a sketch of the proof.

Suppose that a permit market equilibrium $(\{\hat{c}_i\}_{i=1,\ldots,I}, \{\hat{y}_i\}_{i=1,\ldots,I}, \{\hat{\pi}\}_{i=1,\ldots,I}, \hat{p}_e)$ exists. If this equilibrium allocation is Pareto efficient, we must have $MC_i = MC_j$, which is true only if $MRS_i = MRS_j$ because $MC_i = p_e + MRS_i$.

Now let us go back to the Pareto-efficient frontier. By lemma 4 we know that conditions $MC_i = MC_j$ and $MRS_i = MRS_j$ would be met simultaneously by only one PE allocation $(\{c_i^*\}_{i=1,\ldots,I}, \{y_i^*\}_{i=1,\ldots,I}, \{\pi_i^*\}_{i=1,\ldots,I}, a^*)$, which means that any equilibrium that leads to an efficient allocation must have the initial abatement assignments $\{\tilde{a}_i\}$ sum up to a^*.

We are left to show that the equilibrium $(\{\hat{c}_i\}_{i=1,\ldots,I}, \{\hat{y}_i\}_{i=1,\ldots,I}, \{\hat{\pi}_i\}_{i=1,\ldots,I}, \hat{p}_e)$ does exist. Let $\tilde{a} = a^*$, $\hat{\pi}_i = \pi_i^*$, and $\hat{c}_i = c_i^*$. Clearly, $\{\hat{y}_i\}$ and $\hat{p}$ are uniquely determined by

$$\hat{a}_i = a_i^* = \Phi_i(\tilde{Y}_i - \hat{y}_i) = \Phi_i(\bar{Y}_i - y_i^*)$$

and

$$\hat{p}_e = MC_i(\hat{a}_i) - MRS_i(\hat{a}, \hat{c}_i) = MC_i(a_i^*) - MRS_i(a^*, c_i^*) = p_e^*.$$

Finally, the initial assignment of abatement levels is also unique because

$$\begin{aligned}\hat{\theta}_i &= \hat{\pi}_i - \frac{(\hat{c}_i - \hat{y}_i)}{\hat{p}_e \hat{a}} \\ &= \left[\theta_i^* + \frac{(c_i^* - y_i^*)}{p_e^* a^*}\right] - \frac{c_i^* - (\tilde{Y}_i - \bar{Y}_i + y_i^*)}{p_e^* a^*} = \theta_i^* + \frac{\tilde{Y}_i - \bar{Y}_i}{p_e^* a^*}.\end{aligned}$$

Rewrite the last equality to get $(\hat{\theta}_i - \theta_i^*) = (\tilde{Y}_i - \bar{Y}_i)/p_e^* a^*$. The proof is therefore complete. ■

There has been a suggestion that careful distribution of initial permits plus side payments might lead us to an abatement target preferred by a policymaking body. Proposition 6 should convince us that is not possible. The allocation of initial abatement and initial income transfers are two instruments that cannot be separated if a Pareto-efficient allocation of resource is to be achieved. The proposition shows that under the new expanded policy space it is still the case that efficiency would prevent any involvement of a policymaker in either choosing an abatement target or redistributing the initial wealth.

5.7 Lindahl-Pricing Equilibrium

A Lindahl equilibrium is a resource allocation and a set of individual prices $(\{c_i^*\}, \{y_i^*\}, \{\pi_i^*\}, \{p_i^*\})$ that maximizes

$$u_i(c_i, a) \quad \text{subject to} \quad a_i = \Phi_i(\bar{Y}_i - y_i),\ c_i - y_i = (a_i - \bar{a}_i)p_i \qquad (5.17)$$

for each country.

PROPOSITION 7 For any Pareto-efficient abatement allocation $(\{c_i^*\}, \{y_i^*\}, \{\pi_i^*\}, a^*)$, there is a unique allocation of initial abatement levels, such that with these as initial endowments the Lindahl-pricing equilibrium is Pareto efficient.

PROOF. The first-order conditions to problem (5.17) are

$$MC_i(a_i) = p_i + MRS_i(a, c_i). \qquad (5.18)$$

The argument of the proof is basically the following. Let $(\{c_i^*\}, \{y_i^*\}, \{\pi_i^*\}, \{p_i^*\})$ be an arbitrary Pareto-efficient allocation. At this allocation marginal cost, $MC_i(a_i^*)$, and marginal rate of substitution, $MRS_i(a^*, c_i^*)$, for each country are known. Then equation (5.18) uniquely defines country-specific price p_i^*. Finally, the initial abatement assignments $\bar{a}_i$ are determined uniquely by the budget equation, given a_i^*, y_i^*, c_i^*, and p_i^*. The allocation and prices thus constructed, $(\{c_i^*\}, \{y_i^*\}, \{\pi_i^*\}, \{p_i^*\})$, consist of a Lindahl equilibrium.

■

Notice that this proposition says that there is a unique relationship between a Pareto-efficient allocation and the initial abatement assignments that results in that Pareto-efficient allocation as a Lindahl equilibrium. Because in general there is an infinite number of Pareto-efficient allocations $(\{c_i^*\}, \{y_i^*\}, \{\pi_i^*\}, a^*)$ pointing to the same efficient abatement level a^*, there are as many ways of assigning initial abatement levels (with the sum a^*) that are compatible with Pareto efficiency using Lindahl markets with personalized prices.

5.8 An Extension: *M* Private Goods

The extension from one private good to *M* private goods is not a simple matter. Lemma 2, which was used in the proofs of almost all previous results, is not valid anymore. Remember in the case of one private good that production efficiency conditions alone determine the unique relationship between a Pareto-efficient total abatement level $\bar{a}$ and countries' production levels y_i; this is so because we have $(I - 1)$ independent equations $MC_i[a_i(y_i)] = MC_j[a_j(y_j)]$, $\forall\, i \neq j$, and $(I - 1)$ unknowns[11] $(y^1, y^2, ..., y_{I-1})$. With M private goods this is not true anymore: Production efficiency conditions $MC_{i,l}[a_i(\{y_{i,l}\})] = MC_{j,l}[a_j(\{y_{j,l}\})]$, $\forall\, i \neq j$, $\forall\, l$, offer $(I - 1)M$ independent equations, but we have $(IM - 1)$ unknowns; therefore, for $I, M \geq 2$, $y_{i,l}$ could not be determined in the same way as in the case of one private good.

Also the addition of more private goods prevents us from giving a simple proof of the existence of competitive equilibrium. These difficulties force us to take a different approach. In a reversal of the previous approach, we study the Nash equilibrium first and show that under certain regularity conditions there exists a finite number of initial abatement levels, for each of which there exists a finite number of ways of distributing the permits, such that the Nash equilibria under those initial arrangements lead to Pareto-efficient outcomes. We next show that the regularity condition for the existence of a Nash equilibrium is also sufficient for the existence of competitive equilibria. Knowing the existence of an equilibrium, propositions 1 and 2 are then revised for M private goods.

For easy reference we produce the M-private-goods version of Pareto-efficiency conditions. Basically, we have to solve problem (5.4) again. The two constraints to the optimization problem remain the same, except that this time we have to replace the scalars c_i and y_i by vectors $c_i = (c_{i,1}, c_{i,2}, ..., c_{i,M})$ and $y_i = (y_{i,1}, y_{i,2}, ..., y_{i,M})$.

Without difficulty, we arrive at the following necessary conditions for Pareto efficiency:

$$MC_{i,l}(a_i) = MC_{j,l}(a_j), \qquad \forall\, i \neq j \;\text{ and }\; \forall\, l, \tag{5.19}$$

$$\frac{MRS_{i,l}(c_i,\, a)}{MRS_{i,k}(c_i,\, a)} = \frac{MRS_{j,l}(c_j,\, a)}{MRS_{j,k}(c_j,\, a)} \quad \text{for } \forall\, i \neq j,\ \forall\, l \neq k, \tag{5.20}$$

$$\sum_{i=1}^{I} MRS_{i,l}(a,\, c_i) = MC_{j,l}(a_j) \quad \text{for any } j \text{ and } \forall\, l. \tag{5.21}$$

These conditions are essentially the same as conditions for the case of one private good other than the addition of the new condition (5.20), which equalizes marginal rates of substitution between the M private goods across countries.

---

[11] If $\bar{a}$ and $a_1, a_2, ..., a_{I-1}$ are known, then a_I^* is automatically determined.

Nash Equilibrium

With proper modification and derivations, the M-private-good version of the first-order conditions to problem (5.9) is found to be:

$$MC_{i,l}(a_i) = \frac{p_e}{p_l} + MRS_{i,l}(c_i, \bar{a}), \qquad \forall\, i \neq j \quad \text{and} \quad \forall\, l, \tag{5.22}$$

$$\frac{MRS_{i,l}(c_i, \bar{a})}{MRS_{i,k}(c_i, \bar{a})} = \frac{p_e/p_l}{p_e/p_k} \qquad \forall\, i, \forall\, k \neq l. \tag{5.23}$$

Again, a new condition (5.23) is added, stating that the marginal rates of substitution between private consumption goods must equal their market price ratio. The M-private-good version of budget constraint (5.8) now becomes

$$\sum_{l=1}^{M} (c_{i,l} - y_{i,l})p_l = (a_i - \bar{a}_i)p_e, \qquad \forall\, i. \tag{5.8$'$}$$

Rewrite market clearance condition (5.9):

$$\sum_{i=1}^{I} c_{i,l} = \sum_{i=1}^{I} y_{i,l}, \qquad \forall\, l. \tag{5.9$'$}$$

Denote by p the relative price vector $(p_1/p_e, p_2/p_2, ..., p_l/p_e)$. We have the following lemma.

LEMMA 5 Assume that a Nash equilibrium (c_i^*, y_i^*, p^*) exists. The following two conditions are necessary and sufficient for the equilibrium to achieve Pareto efficiency:

(i) $\sum_{i=1}^{I} MRS_{i,k}(c_i^*, \bar{a}) = MC_{j,k}(a_i^*)$ for some j and k,

(ii) $MRS_{i,l}(c_i^*, \bar{a}) = MRS_{j,l}(c_j^*, \bar{a})$ for some l and $\forall\, i \neq j$.

PROOF. *Sufficiency.* By (5.23), clearly Pareto-efficiency condition (5.20) is always satisfied at the equilibrium. Next we show that (5.22), (5.23), and condition (ii) of lemma 5 imply the Pareto-efficiency condition (5.19). From (5.22) and (5.23) we have

$$\frac{MC_{i,l}(a_i)}{MC_{i,k}(a_i)} = \frac{p_e/p_l}{p_e/p_k} = \frac{MRS_{i,l}(\bar{a}, c_i^*)}{MRS_{i,k}(\bar{a}, c_i^*)}$$
$$= \frac{MC_{j,l}(a_i)}{MC_{j,k}(a_i)}, \qquad \forall\, i \neq j \quad \text{and} \quad \forall \text{l}. \tag{5.24}$$

Therefore, if $MC_{i,l}(a_i^*) = MC_{j,l}(a_j^*)$ holds for $\forall\, i \neq j$ and for some l, then it holds for all $l = 1, ..., M$. However, this is exactly what we can get from (5.22) and condition (ii) of lemma 5:

$$\frac{MC_{i,l}(a_i)}{MC_{j,l}(a_i)} = \frac{MRS_{i,l}(\overline{a},\, c_i^*) + \dfrac{p_e}{p_l}}{MRS_{j,l}(\overline{a},\, c_j^*) + \dfrac{p_e}{p_l}} \quad \text{for some } l \quad \text{and} \quad \forall\, i \neq j.$$

Finally, we need to show that (5.21) is also satisfied. This is trivial.

Necessity

We show that if any part of conditions (i) or (ii) does not hold, then the equilibrium cannot achieve Pareto efficiency. It is obvious that if (i) does not hold, then Pareto-efficiency condition (5.21) will be violated. Now consider condition (ii). Suppose that the Nash equilibrium does achieve Pareto efficiency even though $MRS_{i,l}(c^*, \overline{a}) \neq MRS_{j,l}(c^*_{j,l}, \overline{a})$ for some l and some $\forall\, i \neq j$. This immediately leads to contradiction. By (5.22), if $MRS_{i,l}(c^*, \overline{a}) \neq MRS_{j,l}(c^*_{j,l}, \overline{a})$ for some l and some $\forall\, i \neq j$, then $MC_{i,l}(c^*, \overline{a}) \neq MC_{j,l}(c^*_{j,l}, \overline{a})$ for some l and some $\forall\, i \neq j$ contradicting (5.19). ■

We next show that there exist abatement levels such that with proper distribution of those levels as initial abatement endowments Nash equilibria achieve Pareto efficiency. Consider an initial abatement target $\overline{a}$ and an allocation rule $\{\theta_i\}$. According to lemma 5, a Nash equilibrium (c_i^*, y_i^*, p^*), if it exists, achieves Pareto efficiency if the initial $\overline{a}$ and $\{\theta_i\}_{i=1,...,I}$ are properly chosen such that conditions (5.22) to (5.24) and conditions (i) and (ii) of lemma 5 are simultaneously met. These conditions consist of $2(I \times M) + I + M$ independent equations. Note that we have the same number of unknowns, $\{c_{i,l}\}$, $\{y_{i,l}\}$, $\{\overline{a}_i\}$, and $\{p_e/p_l\}$. Denote the $2(I \times M) + I + M$ unknowns by vector x and denote the previous mapping from $\Re^{2(I\times M)+I+M} \rightarrow \Re^{2(I\times M)+I+M}$ by function Ψ. The following regularity condition on Ψ is assumed following a similar assumption in CHS.

Regularity condition. The matrix of first partial derivatives of the function Ψ has full rank.

We now provide the M-private-good version of proposition 3.

PROPOSITION 3′ Assume the *regularity condition.* At no more than a finite number of total abatement levels, and with at most a finite number of ways of distributing them among the countries as their initial endowments, could Nash equilibria lead to Pareto-efficient allocations.

PROOF. The proof part is actually easy. We need only show that the number of equilibrium points to the equation system $\Psi(x) = 0$ is finite. Because the mapping $\Psi : \Re^{2(I\times M)+I+M} \rightarrow \Re^{2(I\times M)+I+M}$ is defined on a compact set in $\Re^{2(I\times M)+I+M}$ and by regularity condition the matrix of first partial derivatives of the function Ψ has full rank, equation system $\Psi(x) = 0$ has at most a finite number of solutions x. This proves proposition 3′. ■

Competitive equilibrium

The extension from one private good to M private goods for the case of Nash equilibrium is the difficult part of this section. The rest becomes easy. Again reproduce the two first-order conditions for competitive equilibrium in the permit market:

$$MC_{i,l}(a_i) = \frac{p_e}{p_l} \qquad \forall\, i,\ \forall\, l, \tag{5.25}$$

$$\frac{MRS_{i,l}}{MRS_{i,k}} = \frac{p_e/p_l}{p_e/p_k} \qquad \forall\, i,\ \forall\, k \neq l. \tag{5.26}$$

The following lemma should be compared with lemma 3.

LEMMA 3′ Assume the regularity condition. Further assume that $\bar{a}$ and $\{\theta_i\}_{i=1,\ldots,I}$ are given. There exists at most a finite number of competitive equilibria $[c_i^*(\{\theta_i\},\bar{a}], y_i^*[\{\theta_i\},\bar{a}], p^*[\{\theta_i\},\bar{a})]$.

PROOF. The logic of this proof is the same as the one to proposition 3′. Here we have unknowns $c_i^*(\{\theta_i\},\bar{a})$, $y_i^*(\{\theta_i\},\bar{a})$, $p^*(\{\theta_i\},\bar{a})$, a total of $[2(I \times M) + M]$. How many equations do we have? The same number: condition (5.25) consists of $I \times M$ equations, (5.26) consists of $I \times (M - 1)$, and (5.24) and (5.25) provide additional $I + M$ equations. Denote the $[2(I \times M) + M]$ unknowns by z and the mapping from $\Re^{2(I\times M)+M}$ to $\Re^{2(I\times M)+M}$ by $\Gamma(z)$. Note that the matrix of first partial derivatives of the function Γ is a submatrix of Ψ. Therefore, the regularity condition on Ψ

implies that the matrix of first partial derivatives of the function $\Gamma(z)$ also has full rank. This ends the proof. ■

Next we show that proposition 1, with small modifications, holds for M private goods. Once again let us inspect the two sets of necessary conditions. We see that the permits market equilibrium conditions (5.25) and (5.26) imply Pareto-efficiency conditions (5.19) and (5.20). However, as before, the Pareto-efficiency condition (5.21) generally will not be automatically satisfied by a competitive equilibrium allocation.

PROPOSITION 1′ Assume the *regularity condition,* and further assume that $\bar{a}$ is given. A distribution of initial abatement $\{\theta_i\}_{i=1,\ldots,I}$ associated with $\bar{a}$ leads to a Pareto-efficient allocation of resources at an equilibrium if and only if that equilibrium allocation $[c_i^*(\{\theta_i\}, \bar{a}), y_i^*(\{\theta_i\}, \bar{a})]$ satisfies condition (i) of lemma 5, or

$$\sum_{i=1}^{I} MRS_{i,k}(c_{i,k}^*, \bar{a}) = MC_{j,k}(a_i^*) \quad \text{for some } j \text{ and some } k. \tag{5.27}$$

PROOF. An equilibrium must satisfy (5.25) and (5.26). Basically, we need to show that if an equilibrium further satisfies (5.27), then the Pareto-efficiency condition (5.21) is also satisfied. From equations (5.25) and (5.26) we have

$$\frac{MRS_{i,k'}}{MRS_{i,k}} = \frac{p_k}{p_{k'}} = \frac{MC_{i,k'}}{MC_{i,k}} \quad \text{for all k}' \neq k \text{ and all } i,$$

from which we easily have

$$\sum_{i=1}^{I} MRS_{i,k'} = \sum_{i=1}^{I}\left(MRS_{i,k}\frac{p_k}{p_{k'}}\right) = \frac{MC_{j,k'}}{MC_{j,k}}\sum_{i=1}^{I} MRS_{i,k} \quad \text{for any } j \text{ and } k' \neq k,$$

or

$$\frac{\sum_{i=1}^{I} MRS_{i,k}}{MC_{j,k}} = \frac{\sum_{i=1}^{I} MRS_{i,k'}}{MC_{j,k'}} \quad \text{for any } j \text{ and any } k' \neq k.$$

Clearly, if $\Sigma_{i=1}^{I} MRS_{i,k} = MC_{j,k}$, then for any $k' \neq k$ we also have $\Sigma_{i=1}^{I} MRS_{i,k'} = MC_{j,k'}$. This completes the proof. ■

There is one crucial point to be made concerning the issue of multiple equilibria. As lemma 3′ states, each initial $\overline{a}$ and its distribution $\{\theta_i\}$ is associated with a finite number of equilibria. We do not know in advance which equilibrium will be realized. It is of course sufficient if all the equilibrium allocations associated with the pair $(\overline{a}, \{\theta_i\})$ satisfy condition (5.27). In that case the total number of constraints on the distributions $\{\theta_i\}$ would be the number of equilibria plus one. The additional constraint, of course, comes from $\Sigma_i \theta_i = 1$. Because we do not know how many equilibria are associated with each $(\overline{a}, \{\theta_i\})$, we are unable to make a definite statement about what $\{\theta_i\}$ would lead to a Pareto-efficient allocation of resources. For the special case of unique equilibrium, the number of constraints on the initial distribution of $\{\theta_i\}$ would be two: one from (5.27) and the other $\Sigma_i \theta_i = 1$.

We have therefore confirmed that proposition 1 indeed holds whether we have one or *M* private goods.

5.9 Concluding Remarks

A global emission permits market has been favored as one of the mechanisms that can effectively control greenhouse gas emissions. A practical implementation issue for such a system to "succeed" is how the emission permits will be allocated between participating countries. Equity is a central issue here. In addition, economists generally would not consider a system successful unless it is efficient. Efficiency and equity are therefore at the center of the debate as to whether a global permit market is preferable to other mechanisms. It has been shown in CHS that the two are inseparable. Here we further refine the CHS findings. We provide necessary and sufficient conditions on the distribution of initial permits such that under those conditions the resulting market equilibrium attains a Pareto-efficient outcome. Furthermore, we extend the CHS model to allow for strategic behavior. We show under this new behavioral assumption on the part of emission abatement participants that the choices of a policymaking body are very limited: At no more than a finite number of total abatement levels and for each abatement level with at most a finite number of ways of distributing the permits among the countries as their initial endowments could the resulting Nash equilibrium lead to Pareto-efficient allocations.

Our equilibrium analysis of countries' initial abatement assignments and permits trading positions provide some topics for further debate or justification. For example, why should it be that, between two countries with the same initial endowments in private goods and the same abatement technology, the country that is more environment conscious is assigned a lower abatement level? Is our efficiency justification sufficient to justify for it? To answer these questions, further research is required.

The model presented here is a static one. We hope to extend it to a dynamic framework in the future. Some preliminary work with two time periods has already be done by Eyckmans, Proost, and Schokkaert [11].

References

1. Chichilnisky, G. "The Abatement of Carbon Emissions in Industrial and Developing Countries." Paper presented at the International Conference on the Economics of Climate Change, OECD/IEA, Paris, 1993. Published in *Economic Approaches to Climate Change,* ed. T. Jones. Paris: OECD, 1994.
2. Chichilnisky, G., and G. M. Heal. "Implementing the Rio Targets: Perspectives on Market-Based Approaches." Working paper, Columbia Business School, Columbia University, 1993.
3. Chichilnisky, G., and G. M. Heal. "Global Environmental Risks." *Journal of Economic Perspectives* 7, no. 4 (fall 1993): 65–86.
4. Chichilnisky, G., and G. M. Heal. "Who Should Abate Carbon Emissions? An International Perspective." *Economics Letters* 44 (1994): 443–49.
5. Chichilnisky, G., and G. M. Heal. "Markets for Tradable CO_2 Emission Quotas: Principles and Practice." Economics Department Working Papers No. 153, OECD, Paris, 1995.
6. Chichilnisky, G., G. M. Heal, and D. A. Starrett. "International Emission Permits: Equity and Efficiency." Paper presented at Conference on Market Approaches to Environmental Protection, Stanford University, 1993.
7. Cline, W. R. "Scientific Basis for the Greenhouse Effect." *Economic Journal* 101 (July 1991): 904–19.
8. Cline, W. R. *The Economics of Global Warming*. Washington, D.C.: Institute for International Economics, 1992.
9. Coase, R. "The Problem of Social Cost." *Journal of Law and Economics* 3 (1960): 1–44.
10. Dales, J. H. *Pollution, Property and Prices.* Toronto: University of Toronto Press, 1968.
11. Eyckmans, J., S. Proost, and E. Schokkaert. "Efficiency and Distribution in Greenhouse Negotiations." *Kyklos* 46 (1993): 363–97.
12. Grubb, M. "Options for an International Agreement." In *Combating Global Warming: Study on a Global System of Tradable Carbon Emission Entitlements.* chap. 2. New York: United Nations, 1992.

13. Hahn, R. W. "Promoting Efficiency and Equity through Institutional Design." *Policy Sciences* 21 (1988): 41–66.
14. Hahn, R. W. "Economic Prescriptions for Environmental Problems: How the Patient Followed the Doctor's Orders." *Journal of Economic Perspectives* 3, no. 2 (1989): 95–114.
15. Hahn, R., and R. Noll. "Designing a Market for Tradable Emission Permits." In *Reform of Environmental Regulation,* ed. W. Magat. Cambridge, Mass.: Ballinger.
16. Hahn, R., and R. Noll. "Barriers to Implementing Tradable Air Pollution Permits: Problems of Regulatory Interactions." *Yale Journal of Regulation* 1 (1983): 63–91.
17. Hahn, R., and R. Noll. "Environmental Markets in the Year 2000." *Journal of Risk and Uncertainty* 3 (1990): 351–67.
18. Hoel, M. "Efficient International Agreements for Reducing Emissions of CO_2." *Energy Journal* 12 (1991): 2.
19. Hoel, M. "International Coordination of Environmental Taxes." *Economics Energy Environment,* Nota Di Lavoro 41.94, 1994.
20. Intergovernmental Panel on Climate Change. *The IPCC First Assessment.* Geneva: World Meteorological Organization/United Nations Environment Programme, 1990.
21. Laffont, J.-J., and J. Tirole. "Environmental Policy, Compliance and Innovation." *Economics Energy Environment,* Nota Di Lavoro 78.93, 1993.
22. Lin, Y. "Exhaustible Resource Extraction, Knowledge Accumulation and Economic Growth." Mimeograph, Columbia University, 1996.
23. Montgomery, W. D. "Markets in Licenses and Efficient Pollution Control Programs." *Journal of Economic Theory* 5, no. 3 (1972): 395–418.
24. Prat, A. "Efficiency Properties of a Constant-Ratio Mechanism for the Distribution of Tradable Emission Permits." Mimeograph, Stanford University, 1995. (Chapter 6 of this volume)
25. Tietenberg, T. "Transferable Discharge Permits and the Control of Stationary Source Air Pollution." *Land Economics* 5 (1980): 391–416.
26. Tietenberg, T. *Emission Trading: An Exercise in Reforming Pollution Policy.* Washington, D.C.: Resources for the Future, 1985.
27. Tietenberg, T. *Environmental and Natural Resource Economics,* 3rd ed. New York: HarperCollins, 1993.

Chapter 6

Efficiency Properties of a Constant-Ratio Mechanism for the Distribution of Tradable Emission Permits

Andrea Prat

6.1 Introduction

The world's public opinion has been increasingly alarmed by the dangers posed by carbon dioxide (CO_2) emissions. The current level of emissions, if not curbed, could lead to relevant climate changes that might have disastrous effects on humanity. Chichilnisky [3] and Chichilnisky and Heal [4] offer a general review of the problem of CO_2 emissions. Such a complex issue can be analyzed from several viewpoints. This chapter focuses on the public good aspect. As CO_2 tends to distribute itself evenly in the atmosphere over time, in the long run it does not matter where on the earth's surface CO_2 originates; what matters is only the global amount of emissions. Carbon dioxide closely approximates a global public good.

To curb or at least slow the growth of CO_2 emissions, a mechanism needs to be devised to deal with the public good problem. Two possibilities are direct regulation and discouraging taxation. A third possibility, which forms the object of this chapter, follows the Coasian tradition and consists of distributing tradable emission permits.

I am grateful to Kenneth Arrow, Graciela Chichilnisky, Geoffrey Heal, Michael Smart, Valter Sorana, and David Starrett for their helpful comments.

In the simplest version a tradable emission permit mechanism would work as follows. An international market for emission permits is set up. Each country receives a given amount of emission permits. If a country pollutes more than its amount of permits allows for, it should make up for the difference by buying permits on the international market. If it pollutes less, it can sell the unused permits. It is common wisdom that such a mechanism would bring about production efficiency: Countries will face a powerful incentive to develop and apply low-pollution technologies.

However, Chichilnisky, Heal, and Starrett [5] examine the problem of Pareto efficiency for a tradable emission permit mechanism and prove that, given a global level of emissions and a distribution of tradable permits, the competitive equilibrium allocation is, in general, not Pareto efficient. To reach a Pareto-efficient allocation, the planner needs to look for some special permit allocation.

This chapter takes a different perspective on the same problem. Instead of holding the global level of emissions constant and looking for the "right" distribution, the reverse is done; that is, given some exogenous ratios, the aim is to find a global level of emissions that gives a Pareto-efficient allocation. The main proposition is that, given regularity conditions, that level exists and is unique.

A constant-ratio mechanism has three logical stages. First, each country is exogenously entitled to a constant ratio of all emission permits that will be issued. Second, the planner chooses the total amount of emissions. Third, each country receives its share of permits and is free to trade them for consumption goods. A constant-ratio mechanism can be seen as a way to separate the distribution issue from the efficiency issue.

The starting point of a constant-ratio mechanism is the definition of property rights over a special factor of production, emissions, which is similar to the definition of property rights over other factors, say, offshore oil. One possibility is to define such property rights as a set of ratios of any future emission that each country is entitled to.[1] For instance, one country could be entitled to 10% of all the world's emission permits whatever the global level of emissions will be. An entitlement to a constant ratio of emission permits has an analogy to a property right over a corresponding fraction of the atmosphere. A country entitled to 10% of all the world's permits could be viewed as the owner of 10% of the atmosphere. Obviously, this property right is incomplete,

[1] For instance, candidates for proportions could be population shares or current emission shares.

as a country cannot decide independently the level of pollution of its share of the atmosphere.

The mechanism lets countries trade their permits. A country can pollute more than its share allows for by buying permits from another country or, conversely, can pollute less than its share and sell part of its permits. Then, as is easy to see, the marginal productivity of emissions will be equalized to the international price of emission rights in all countries. Permit trade alone guarantees efficiency on the production side.

However, as Chichilnisky, Heal, and Starrett have shown in chapter 3 of this volume, if we look at the consumption side, we run into the public good problem, and we see that, in general, countries are not satisfied with the allocation that results from a given level of emissions through a competitive equilibrium. Suppose, for instance, that an overwhelming majority of countries want to decrease the current level of emissions while only a minority want to keep it constant or to increase it (of course, each country has taken into account the effect that a decrease will produce on its utility both directly as a decrease of a public bad and indirectly through a decrease in consumption due to less available production factor). Then, if side transfers are possible, there exists an alternative allocation at which the emission level is decreased and the majority of countries that benefit from the decrease compensate with consumption goods the minority that are hurt. Such an alternative allocation is Pareto improving. Therefore, to be efficient, a level of emissions needs to be resistant to recontracting among countries. In an intuitive sense the global emission level must be such that the thrust of the countries that want to increase it exactly offset the thrust of the countries that want to decrease it. In this chapter such a concept is formalized by a marginal willingness-to-pay function.

Section 6.2 contains the main propositions. It is shown that, in the constant-ratio mechanism, for each vector of ratios, there exists a unique global level of emissions that results in a Pareto-efficient allocation. Pareto efficiency is defined in the broadest sense. In the hope of making the exposition more intuitive, the proof is given for a world with N countries, one private good and one factor of production (emissions). The Appendix generalizes the result to a model with several private goods and several production factors, both traded and non-traded.

Section 6.3 touches the issue of implementation. Once it is established that a constant-ratio mechanism can reach a Pareto-efficient allocation, the question is, If countries vote on the global level of emissions, what will happen? It turns out that there exists a unique voting equilibrium at which countries vote in a straightforward manner but that this equilibrium need not coincide with the Pareto-efficient level of emissions.

6.2 Pareto Efficiency

In this deterministic[2] model there is one homogeneous consumption good, c. To produce it, it is necessary to produce polluting emissions. Emissions can be regarded as a production factor for c. Given a technology, if we want to produce more c, we need to pollute more. In this simplified model emissions will be the only argument of the pollution function. Utility depends on two arguments: the consumption of private good and the consumption of the public bad.

There are N countries,[3] each of which has a country-specific utility function $U_i(c_i, e)$—the arguments are the country's private consumption and the world's level of emissions—and a country-specific production function $f_i(e_i)$—the argument is the amount of emission used by the country to produce private goods. The production function is strictly concave. The utility function is increasing in c, decreasing in e, twice-continuously differentiable, and strictly quasi convex. Moreover, the marginal rate of substitution between consumption and pollution $U^i_e(c, e)/U^i_c(c, e)$ is assumed to be strictly decreasing in c and in e (i.e., $U^i_{-e}(c, -e)/U^i_c(c, -e)$ is strictly decreasing in $-e$ air quality and strictly increasing in c: both air quality and the consumption good are normal goods).[4]

Also,[5]

$$\lim_{e_i \to 0^+} f'_i(e_i) = \infty \quad \text{and} \quad \lim_{e_i \to \infty} f'_i(e_i) = 0 \qquad i = 1, 2, ..., N.$$

A constant-ratio mechanism for the allocation of emission permits determines each country's amount of permits ϵ_i as follows:

$$\epsilon_i = \pi_i e$$

---

[2]The double uncertainty in the connection between CO_2 emissions and global heating and between global heating and effects on human activity is a fundamental feature of the global warming issues and poses a series of problems in a dynamic context. This model is both static and deterministic.

[3]In principle, the constant-ratio mechanism should be based on people and firms and not on countries. People would be entitled to shares of the world's emission amount, which they would sell to firms. Firms would produce using permits bought from people. In this chapter, the word *agent* (be it a consumer or a producer) could as well replace the word *country*. However, all the current discussions focus on the role of countries. Therefore, this model will be based on countries with all the caveats that aggregate utility functions entail.

[4]This assumption is used to prove uniqueness but is not needed for existence.

[5]Assumption

$$\lim_{e_i \to \infty} f'_i(e_i) = 0$$

can be replaced with

$$\lim_{e_i \to \infty} f_i{}'(e_i) = 0 \quad \text{with a} \in (\infty, 0].$$

with

$$\pi \geq 0, \qquad i = 1, 2, ..., N \sum$$

$$\sum_{i=1}^{N} \pi_i = 1.$$

The ratios π_i are predetermined and are held constant. The countries are free to trade their share of emission permits.

In this model two factors determine a country's level of consumption: its technology and its ratio of permits. A country with an efficient technology will produce more goods. At the margin using a permit to produce goods or selling the permit for consumption good is equivalent. However, efficient countries can earn a larger surplus before they get to the margin. The second source of difference is the ratio of permits. A country with a high share of permits will either sell them for consumption good or use them to produce without the need of buying permits from other countries. In the general model, treated in the Appendix, the differences between countries will also depend on the endowments of factors of production.

Finally, the model includes a planner, whose only decision variable is the total level of emissions

$$e \in [0, \infty).$$

DEFINITION An allocation $(e; e_1, ..., e_N; c_1, ..., c_N)$ is Pareto efficient in an unrestricted sense if there does not exist a different allocation, that may involve side transfers in consumption good, that makes no country worse off and at least one country better off.

DEFINITION At a given e, a competitive equilibrium is given by

$$< c_1^*(e), ..., c_N^*(e); e_1^*(e), ..., e_N^*(e); p(e) >$$

that satisfy, for $i = 1, 2, ..., N$,

$$V_i(e) = \max Ui(ci, e) \quad \text{subject to} \quad c_i - f_i(e_i) = p(\pi_i e - e_i), \qquad e_i \geq 0$$

and

$$\sum_{i=1}^{N} e_i = e.$$

The term $V_i(e)$ is the maximized utility function for country i and depends on e and on all the π's.

Notice that, for any global level of emissions $e \in [0, \infty)$, the necessary conditions for competitive equilibrium correspond to production efficiency:

$$f'_i(e^*_i(e)) = p \quad \text{for } i = 1, 2, ..., N$$

LEMMA 1 Given e, there exists a unique competitive equilibrium $(c^*_1(e), ..., c^*_N(e); e^*_1(e), ..., e^*_N(e); p(e))$.

PROOF. When e is held constant, this model has one good (c), one factor of production (e), N producers, and N consumers (consumer i is entitled to all the profits of producer i and none of the profits of producers $j \quad i$).

The assumptions that $f_i{}' < 0$ and that

$$\lim_{e_i \to 0^+} f'_i(e_i) = \infty \quad \text{and} \quad \lim_{e_i \to \infty} f'_i(e_i) = 0 \qquad i = 1, 2, ..., N$$

ensure that the solution to $f_i{}'(e_i) = p$ exists and is unique in all countries. Therefore, the solution to the equation

$$\sum_{i=1}^{N} e^*_i(p) = e$$

exists and is unique. As the $c^*_i(e)$ are uniquely determined by the trade balance constraints, it follows that a competitive equilibrium exists and is unique. ■

DEFINITION The marginal willingness-to-pay function for country i is defined as

$$MW_i(e) = \frac{V'_i(e)}{U^i_c(c^*_i(e), e)}.$$

When it is positive (negative), $MW_i(e)$ represents the amount of consumption that good country i is willing to forgo in exchange for a marginal increase (decrease) in the total emission level e.

Let us pause on the interpretation of MWi. By the envelope theorem,

$$\frac{dc^*_i(e)}{de} = \pi_i p(e).$$

Then, for country i,

$$MW_i(e) = \frac{U'_e(c^*_i(e), e)}{U^i_c(c^*_i(e), e)} + \pi_i p(e).$$

The first addend corresponds to the marginal rate of substitution between global emission level and consumption for country i. As U^i_e is negative, the first addend is negative. On the other hand, the second addend is positive and decreases as e increases. At $\hat{e}$,

$$-\frac{U'_e(c^*_i(\hat{e}),\, \hat{e})}{U^i_c(c^*_i(\hat{e}),\, \hat{e})} = \pi_i p(\hat{e}),$$

so that $MW_i(\hat{e})$. Given that vector of ratios, $\hat{e}_i$ is the bliss point for country i. If $e > \hat{e}$, then country i would like to see e decrease and vice versa. In general $\hat{e}_i$ will differ from country to country.

So far we have looked at a single country. If we turn to the aggregate, we can imagine that the efficient e will be such that the pressure from countries who want a higher e equals the pressure from countries who want a lower e. To formalize this concept we will use the notion of marginal willingness-to-pay aggregate function, defined as

$$MW(e) = \sum_{i=1}^{N} MW_i(e).$$

The term $MW(e)$ can be viewed as a general willingness to move e. For instance, if, at e, $MW(e)$, then a new allocation, possibly including side transfers, can be found at $\hat{e} > e$ such that all countries are better off.

LEMMA 2 $MW(e)$ is continuous and strictly decreasing and there exists a unique $\hat{e}$ such that $MW(\hat{e})$.

PROOF. To prove continuity, consider

$$\begin{aligned} MW(e) &= \sum_{i=1}^{N} MW_i(e) \\ &= \sum_{i=1}^{N} \frac{U^i_e(c^*_i(e),\, e)}{U^i_c(c^*_i(e),\, e)} + f'_j(e^*_j(e)) \quad \text{for any } i = 1, 2, ..., N. \end{aligned}$$

By assumption, U^i_e, U^i_c and f'_j are continuous, and $U^i_c > 0$. Therefore, $MW(e)$ is continuous.

To prove the "strictly decreasing" part, it will be proven that both addends are strictly decreasing. First, let us prove that the first addend is decreasing for each country. Recall that the marginal rate of substitution is decreasing in both c and e and that

$$\frac{dc_i^*(e)}{de} = \pi_i p(e).$$

Therefore,

$$\frac{d\dfrac{U_e^i}{U_c^i}}{de} = \frac{\partial\dfrac{U_e^i}{U_c^i}}{\partial e} + \frac{\partial\dfrac{U_e^i}{U_c^i}}{\partial c}\frac{dc_i^*(e)}{de} < 0 \quad \text{for } i = 1, 2, ..., N.$$

Next let us prove that the second addend is decreasing as well. Consider

$$p(e) = f_i'(e_i^*(e))$$

$$\frac{dp}{de} = f'\frac{de_i^*}{de}$$

$$\sum_{i=1}^{N}\frac{de_i^*}{de} = 1$$

$$\sum_{i=1}^{N}\frac{1}{f_i'}\frac{dp}{de} = \frac{dp}{de}\sum_{i=1}^{N}\frac{1}{f_i'} = 1.$$

Because $f_i' < 0$ for all i, then dp/de. Therfore, $MW(e)$ is (strictly) decreasing.

To prove existence and uniqueness, recall that

$$\lim_{e_i \to 0^+} f_i'(e_i) = \infty \quad \text{and} \quad \lim_{e_i \to \infty} f_i'(e_i) = 0 \qquad i = 1, 2, ..., N.$$

Furthermore, as by assumption $U_e^i(0, e)/U_c^i(0, e)$ is negative and decreasing in e, then

$$\lim_{e_i \to 0^+}\frac{U_e^i(0, 0)}{U_c^i(0, 0)}$$

is a bounded negative number. Therefore,

$$\lim_{e \to 0^+} MW(e) = +\infty$$

$$\lim_{e \to \infty} MW(e) \leq 0.$$

Then, because $MW(e)$ is continuous, there exists an $\hat{e}$ such that $MW(\hat{e}) = 0$. As $MW(e)$ is also strictly decreasing, $\hat{e}$ is unique.

The properties of the marginal willingness-to-pay function proven here provide a tool for demonstrating the main result of this chapter.

PROPOSITION 3 In a constant-ratio mechanism there exists a unique global level of emissions $\hat{e}$ that results in a competitive equilibrium corresponding to a Pareto-efficient allocation.

PROOF. To prove the proposition we state the conditions for unrestricted Pareto efficiency and show that there exists a unique level of emissions $\hat{e}$ such that a constant-ratio mechanism allocation satisfies those conditions.

By lemma 1, for a given e, a constant-ratio mechanism results in a unique competitive equilibrium allocation

$$< e_1^*(e),\ ...,\ e_N^*(e);\ c_1^*(e),\ ...,\ c_N^*(e),\ e.$$

Given the convexity of the problem, the first-order conditions for Pareto efficiency are necessary and sufficient. The conditions for unrestricted Pareto efficiency (the planner chooses all the variables) are

$$f_1'(e_1) = f_2'(e_2) = \cdots = f_N'(e_N), \tag{6.1}$$

$$\lambda_1 U_c^1(c_1, e) = \lambda_2 U_c^2(c_2, e) = \cdots = \lambda_N U_c^N(c_N, e), \tag{6.2}$$

and

$$\sum_{i=1}^{N} \lambda_i U_e^i(c_i, e) + f_j'(e_j) = 0 \quad \text{for any } j. \tag{6.3}$$

First, notice that (6.1) is always satisfied by $e_1^*(e), ..., e_N^*(e)$. Now consider all the possible constant-ratio mechanism allocations: $< e_1^*(e), ..., e_N^*(e), c_1^*(e), ..., c_N^*(e), e >$.

CLAIM 1 If $e = \hat{e}$ and $\lambda_i (1/U_c^i(c_i^*(\hat{e}), \hat{e}))$ for $i = 1, 2, ..., N$, then $< e_1^*(e), ..., e_N^*(e), c_1^*(e), ..., c_N^*(e), e >$ satisfy (6.1) to (6.3).

PROOF. Equation (6.1) is always satisfied. Obviously, (6.2) is satisfied. With these λ's, (6.3) coincides with $MW(e) = 0$, which, by lemma 2, is satisfied if $e = \hat{e}$.

CLAIM 2 If $e \neq \hat{e}$ or $\lambda \neq a(1/U_c^i(c_i^*(\hat{e}), \hat{e}))$ for $i = 1, 2, ..., N$ $a \in [0, \infty)$, then $< e_1^*(e), ..., e_N^*(e), c_1^*(e), ..., c_N^*(e), e >$ cannot satisfy (6.1) to (6.3).

PROOF. If $\lambda \neq a(1/U_c^i(c_i^*(\hat{e}), \hat{e}))$, then (6.3) does not hold, and claim 2 is proven. Suppose that $\lambda_i = a(1/U_c^i(c_i^*(\hat{e}), \hat{e}))$. Then (6.3) coincides with $MW(e) = 0$. However, by lemma 2, if $e \neq \hat{e}$, then $MW(e) \neq 0$, and (6.3) does not hold.

Claim 1 proves existence. Claim 2 proves uniqueness. ■

For the sake of exposition, the proof was given for a one-good, one-factor model. However, it is possible to generalize the assumptions of the model. Suppose there is a vector of consumption goods c, a vector of internationally traded factors of production k, and a vector of noninternationally traded factors of production l. Proposition 1 still holds. For the proof, see the Appendix.

6.3 Implementation

So far it has been assumed that a planner is to choose the global level of emissions. Then, given a set of ratios, this planner can always find a Pareto-efficient level. However, the planner needs to know every country's utility function and production set, which is a heavy informational requirement. Is it possible to decentralize the choice of the emission level?

In this section majority voting sets the global emission level.[6] Each country has one vote. Given a level e', another level e'' is proposed, votes are taken, and the level that receives the greater number of votes is implemented. Successive rounds of voting are taken until a global level of emission e^M is reached such that no other e can get a greater number of votes. Such a level e^M is called a voting equilibrium .

As Gibbard [6] showed, in general a unique voting equilibrium need not exist. However, a constant-ratio mechanism yields the following.

PROPOSITION 4 In a constant-ratio mechanism there exists a unique voting equilibrium e^M, where e^M is the global emission level desired by the median voter.

[6]Bowen [2] studied the problem of voting on the level of a public good to be provided through taxation. Citizens share the tax burden equally. Here the problem is analogous. A public good, clean air, is provided through taxation in predefined ratios. The only difference is that whereas in Bowen's model taxation hits a consumption good, here it hits a production factor.

PROOF. Consider the function $V_i(e)$ defined by

$$V_i(e) = \max U^i(c_i, e) \quad \text{subject to } c_i - f(e_i) = p(\pi_i e - e_i), \qquad e_i \geq 0.$$

The term $V_i(e)$ is continuous. By extending lemma 2, for each i, there exists a unique $\hat{e}$ such that $MWi(\hat{e}) = 0$, which implies that there exists a unique $\hat{e}_i$ such that $V_i^i(\hat{e}) = 0$. Then $V_i(e)$ is single peaked for all countries. Then[7] there exists a unique voting equilibrium e^M, where e^M is the global emission level desired by the median voter. ■

In general, the voting equilibrium e^M will be different from the Pareto-efficient level $\hat{e}$. The condition that determines e^M is

$$MW_M(e) = \frac{U_e^M(c_M^*(e), e)}{U_c^M(c_M^*(e), e)} + \pi_M p(e) = 0,$$

where M is the median voter and the condition that determines $\hat{e}$ is

$$MW(e) = \sum_{i=1}^{N} \frac{U_e^M(c_i^*(e), e)}{U_c^M(c_i^*(e), e)} + p(e) = 0.$$

Under some simplifying analytical assumptions, it is possible to state an intuitive condition under which majority voting yields the efficient level.

PROPOSITION 5 If all countries have identical isoelastic utility functions and receive equal ratios of emission permits, then the voting equilibrium e^M and the Pareto-efficient level $\hat{e}$ coincide if and only if the mean income and the median income coincide.

The assumption of identical utility function corresponds to assuming that differences in the way countries value clean air are due only to income differences. If two countries have the same income, they demand the same amount of clean air. This excludes cultural differences, that is, cases in which citizens of some countries might value clean air over consumption intrinsically more than citizens of other countries. Of course, technological differences are still present.

PROOF. Suppose that

$$U_i(c_i, e) = (E - e)^a c_i^b.$$

[7] As proven in Black [1].

Then

$$\frac{U_e^i(c_i^*(e), e)}{U_c^i(c_i^*(e), e)} = \frac{ac_i}{b(E - e)},$$

so that Pareto efficiency implies that

$$\frac{U_e^M(c_M^*(e), e)}{U_c^M(c_M^*(e), e)} + \pi_M p(e) = -\frac{ac_i}{b(E - e)} + \frac{1}{N}p(e) = 0,$$

whereas the voting equilibrium requires that

$$\sum_{i=1}^{N} \frac{U_e^i(c_i^*(e), e)}{U_c^i(c_i^*(e), e)} + p(e) = -\sum_{i=1}^{N} \frac{ac_i}{b(E - e)} + p(e) = 0$$

and

$$-\frac{aE(c)}{b(E - e)} + \frac{1}{N}p(e) = 0,$$

and the two conditions are identical if and only if $E(c) = c_M$. As there are no savings, the voting equilibrium e^M and the Pareto-efficient level $\hat{e}$ coincide if and only if the mean income and the median income coincide. ■

If the income distribution is skewed toward lower incomes, as the world distribution is, then the mean income is higher than the median income. Proposition 3 indicates that the global level of emission achieved through a voting equilibrium will not be Pareto efficient. Given the voting equilibrium, there could be a Pareto-improving alternative allocation whereby developed countries transfer income toward developing countries in exchange for a decrease in the global level of emissions. Therefore, a constant-ratio mechanism, if implemented through voting, is likely to bring about a global emission level that is higher than the one that an omniscient planner would choose.

6.4 Remarks and Conclusions

The result of existence of a Pareto-efficient allocation is very robust. Mainly, it depends on the fact that, if $e = 0$, all countries want e to increase, whereas if e is large enough, all countries want e to decrease. It is easy to see that existence still holds if we take the share of emission permits to be functions instead of constants.

A mechanism for dealing with public goods should have two desirable properties. First, it should separate the issue of efficiency from the issue of equity. Second, it should be implementable with decentralized information.

Regarding the first property, a constant-ratio mechanism is entirely satisfactory. The issue of equity involves selecting a vector of ratios. The fundamental question of the choice of the ratios is outside the scope of this chapter. However, once the vector of ratios is determined, the issue of efficiency can be solved uniquely and no recontracting can make countries better off.

Regarding the second property, a constant-ratio mechanism yields mixed results. On the bright side it has a unique voting equilibrium in which countries vote in a straightforward manner. However, this equilibrium need not coincide with the efficient level. The gap between the two depends on the difference between the zeroes of the marginal aggregate willingness to pay and the median willingness to pay.

Appendix

Suppose there are N countries, M consumption goods c, Q internationally traded production factors k, and P noninternationally traded production factors l. There are MN production functions, one for each country and each good.

Proposition 1 holds.

PROOF. Here the predicate of lemma 1 will be assumed, not derived; namely, it will be assumed that, for each level of e, there exists a unique competitive equilibrium allocation.[8]

Besides the respect of constraints, the conditions for a competititive equilibrium, given e, are

$$p^j \frac{\partial f_i^j}{\partial e_i^j} = p \qquad i = 1, \ldots, N, j = 1, \ldots, M, \tag{A6.1}$$

$$p^j \frac{\partial f_i^j}{\partial k_i^h} = q^h \qquad i = 1, \ldots, N, j = 1, \ldots, M, h = 1, \ldots, Q, \tag{A6.2}$$

$$\frac{\partial f_i^j}{\partial l_i^g} = \chi_i^g \qquad i = 1, \ldots, N, j = 1, \ldots, M, g = 1, \ldots, P, \tag{A6.3}$$

[8]The analysis of the conditions for existence and uniqueness of competitive equilibrium is outside the scope of this chapter. What we want to prove is that, if existence and uniqueness are already there, then a constant-ratio mechanism will preserve them.

$$\frac{\partial U^i}{\partial c_i^j} = \gamma_i p^j \qquad i = 1, ..., N, j = 1, ..., M, \tag{A6.4}$$

$$\sum_{g=1}^{P} \sum_{j=1}^{M} (\bar{l}_i^g - l_i^g) = 0, \tag{A6.5}$$

and

$$\sum_{j=1}^{M} p^j(c_i^j - f_i^j) = p\left(\pi_i e - \sum_{i=1}^{M} e_i^j\right) + \sum_{h=1}^{Q} \sum_{j=1}^{M} q_h(k_i^h - k_i^{hj}) \quad \text{for } i = 1, ..., N, \tag{A6.6}$$

where p is the price of emission permits, $(p^1, p^2, ..., p^M)$ are the price of consumption goods, and $(q^1, q^2, ..., q^Q)$ are the prices of traded factors. The γ's and χ's represent Lagrange multipliers. The first three conditions correspond to efficiency in production, the fourth condition ensures efficiency in consumption bundles, the fifth condition corresponds to the constraints for nontraded resources, and the sixth condition corresponds to the satisfaction of trade balance for each country. A competitive equilibrium determines an allocation (where c^*, e^*, k^*, and l^* are matrices),

$$< c^*(e), e^*(e), k^*(e), l^*(e), e >.$$

Let us take the price of good 1 as numeraire, that is, $p^1 = 1$. The marginal willingness-to-pay function for country i is

$$MW_i(e) = \frac{V_i'(e)}{\frac{\partial U^1}{\partial c_1^i}} = \frac{\frac{\partial U^i}{\partial e} + \gamma_i(e)p(e)\pi_i}{\frac{\partial U^1}{\partial c_1^i}} = \frac{\frac{\partial U^i}{\partial e}}{\frac{\partial U^1}{\partial c_1^i}} + p(e)\pi_i = \frac{\frac{\partial U^i}{\partial e}}{\frac{\partial U^1}{\partial c_1^i}} + \frac{\partial f_i^1}{\partial e_i^1}\pi_i.$$

The marginal willingness-to-pay aggregate function is

$$MW(e) = \sum_{i=1}^{N} \frac{\frac{\partial U^i}{\partial e}}{\frac{\partial U^1}{\partial c_1^i}} + p(e) = \sum_{i=1}^{N} \frac{\frac{\partial U^i}{\partial e}}{\frac{\partial U^1}{\partial c_1^i}} + \frac{\partial f_m^1}{\partial e_m^1} \quad \text{for any } m.$$

The term $MW(e)$ is a scalar and is analogous to the one-good case. It is easy to check that lemma 2 applies and there exists a unique $\hat{e}$ such that $MW(\hat{e}) = 0$.

Now let us replicate the proof of proposition 1. The conditions for unrestricted Pareto efficiency are

$$\text{same as (A6.1–A6.5)} \tag{A6.1$'$}$$

$$\lambda_1 \frac{\partial U^1}{\partial c_1^1} = \lambda_2 \frac{\partial U^2}{\partial c_2^1} = \cdots = \lambda_N \frac{\partial U^N}{\partial c_N^1}, \tag{A6.2$'$}$$

and

$$\sum_{i=1}^{N} \lambda_i \frac{\partial U^i}{\partial e} + \lambda_m \frac{\partial U^m}{\partial c_m^1} \frac{\partial f_n^1}{\partial e_n^1} = 0 \quad \text{for any } m \text{ and for any } n. \tag{A6.3$'$}$$

Of course, in (A6.2′) and (A6.3′) any index j could substitute 1.

If we take $\lambda_i = 1/(\partial U^1/\partial c_i^1)$ for all i, we have

$$\text{(A6.3$'$)} = MW(e) = \sum_{i=1}^{N} \frac{\frac{\partial U^i}{\partial e}}{\frac{\partial U^1}{\partial c_i^1}} + \frac{\partial f_m^1}{\partial e_m^1} \quad \text{for any } m.$$

Then, by the fact that $MW(e)$ has a unique solution $\hat{e}$, it is straightforward to see that there exists a unique case where $< c^*(e), e^*(e), k^*(e), l^*(e), e >$ satisfy (A6.1′) to (A6.3′), that is, when

$$e = \hat{e} \quad \text{and} \quad \lambda_i = \frac{1}{\frac{\partial U^1}{\partial c_i^1}} \quad \text{for } i = 1, 2, ..., N.$$

Proposition 1 holds for the general case. ■

References

1. Black, D. *The Theory of Committees and Elections.* Cambridge: Cambridge University Press, 1958.
2. Bowen, H. "The Interpretation of Voting in the Allocation of Economic Resources." *Quarterly Journal of Economics* 58 (1943): 27–48.
3. Chichilnisky, G. "The Abatement of Carbon Emissions in Industrial and Developing Countries: A Comment." Paper presented at OECD Confer-

ence on "The Economics of Climate Change," OECD Paris, June 14–16, 1993. Published in *OECD: The Economics of Climate Change,* ed. Tom Jones (Paris: OECD, 1994), pp. 159–70.

4. Chichilnisky, G., and G. Heal. "Global Environmental Risks." *Journal of Economic Perspectives* 7, no. 4 (fall 1993): 65–86.
5. Chichilnisky, G., G. Heal, and D. Starrett. "International Emission Permits: Equity and Efficiency." Discussion Paper No. 381, Center for Economic Policy and Research, 1993.
6. Gibbard, A. "Manipulation of Voting Schemes: A General Result." *Econometrica* 42 (1975): 587–601.

Chapter 7

Who Should Abate Carbon Emissions? An International Viewpoint

Graciela Chichilnisky
Geoffrey Heal

7.1 Who Should Abate?

The 1992 Earth Summit in Rio de Janeiro acknowledged the need for international cooperation in responding to the threat of climate change posed by the rapidly increasing concentration of carbon dioxide (CO_2) in the atmosphere. There are, however, substantial differences of opinion both about the main issues and about the framework for resolving them. Industrial countries typically focus on the potential problems posed by the growth of population in developing countries and on the environmental pressure from carbon emissions that this could create over the next half century. Abatement efforts, they feel, should be initiated in the developing countries. On the other hand developing countries view the carbon emission problem as one that originates historically and currently in the industrial countries and that requires their immediate action. Indeed the large majority of all carbon emissions, about 73%, originate currently and historically in the OECD countries and in the ex-Soviet Union; the developing countries have almost four-fifths of the world's population yet contribute at most 30% of all carbon emissions.[1]

Carbon dioxide emissions are a by-product of animal life and of economic activity that involves burning fossil fuels. The rapid increase in the concentra-

Reprinted from *Economics Letters*, vol. 44, 1994, pp. 443–49, Chichilnisky et al.,"Who Should Abate Carbon Emissions?"

[1]There is more detail in Chichilnisky [2–4] and Chichilnisky and Heal [5].

tion of CO_2 in the atmosphere that has occurred since World War II has become a matter of great concern, as it could lead to major and irreversible climate changes. This concentration affects all of us equally because CO_2 mixes uniformly throughout the planet's atmosphere.

From the economic viewpoint, therefore, the abatement of carbon emissions increases our consumption of a public good, a "better" atmosphere. However, this differs from the classic public good in that it is not produced in a centralized fashion. Its production is decentralized: Each consumer of the atmosphere is also a producer. Each country uses the atmosphere as a "sink" for the carbon emissions that are a by-product of its economic activities. We have, therefore, a public good that is independently produced as well as consumed by all, a case that is closer to that of an economy with externalities (e.g., Baumol and Oates [1] and Heal [7]). The classic questions of optimality in the provision of the public good now become questions about the optimal abatement levels of the different countries. Who shall abate, and by how much? How are the optimality conditions for abatement related to the countries' levels of income, their marginal costs of abatement, and the efficiency of their abatement technologies?

We find some answers to these questions in a simple model of the world economy (introduced in Chichilnisky [4]) consisting of a finite number of countries.[2] Each country has a utility function that depends on the consumption of a public good and of a private good, such as income. The production of private good emits CO_2 as a by-product, and in each country the private good can be transformed into the public good through an abatement technology.

We show that Pareto efficiency dictates that the marginal cost of abatement in each country must be inversely related to that country's marginal valuation for the private good (proposition 1). In particular, it is not generally true that Pareto optimality requires that marginal abatement costs be equated across countries. This is true only if marginal utilities of income are equated across countries, either by assumption or by lump-sum transfers across countries. If richer countries have a lower marginal valuation of the private good, then at a Pareto-efficient allocation, they should have a larger marginal cost of abatement than the lower-income countries. With diminishing returns to abatement, this implies that they should push abatement further.

There is a presumption in the literature that efficiency requires equalization of marginal abatement costs. This presumption underlies proposals for the use of uniform carbon taxes and tradable carbon emission permits (Weyant [9] and Coppel [6]). However, in view of the public good nature of the atmosphere and the fact that carbon emissions are produced in a decentralized fashion, effi-

---

[2]It is, in fact, consistent with that of Baumol and Oates [1], chapter 4.

ciency will not in general require the equalization of marginal costs of abatement across countries without lump-sum transfers.

In a two-country example we show that, at an efficient allocation, the quantity of income allocated by a country to abatement is inversely proportional to the level of income—or consumption—of that country, with the constant of proportionality increasing with the efficiency of the country's abatement technology (proposition 2).

The equalization of marginal costs would be necessary for Pareto efficiency if the goods under consideration were private goods. However, in our case we are dealing with a public good, that is, one that, by definition, is consumed by all in the same quantity: the atmospheric CO_2 concentration. This public good is "produced" by the CO_2 emissions (or by the abatement of these emissions) of a finite number of large agents, namely, the countries. In this sense it differs from the classical treatments of Lindahl and Bowen, which were extended subsequently by Samuelson (see Atkinson and Stiglitz [8] p. 489, n. 3). In those cases the public good is produced by a single agent, as is the case for a law and order or defense.

7.2 Pareto-Efficient Abatement Strategies

Consider a world economy with N countries, $N \geq 2$, indexed by $n = 1, ..., N$. Each country has a utility function u_n, which depends on its consumption of private goods, c_n, and on the quality of the world's atmosphere, a, which is a public good. Formally, $u_n(c_n, a)$ measures welfare, where $u_n : R^2 \to R$ is a continuous, concave function and $\partial u_n/\partial c_n > 0$, $\partial u_n/\partial a > 0$. The quality of the atmosphere, a, is measured by, for example, the reciprocal or the negative of its concentration of CO_2. The concentration of CO_2 is "produced" by emissions of carbon, which are positively associated with the levels of consumption of private goods, c_n, that is,

$$a = \sum_{n=1}^{N} a_n, \quad \text{where } a_n = \Phi_n(c_n),$$
$$\text{for each country } n = 1, ..., N, \ \Phi_n' < 0 \forall n. \tag{7.1}$$

The term a is a measure of atmospheric quality overall and a_n is an index of the abatement carried out by country n. The production functions Φ_n are continuous and show the level of abatement or quality of the atmosphere decreasing with the output of consumption. An allocation of consumption and abatement across all countries is a vector

$$(c_1, a_1, \ldots, c_N, a_N) \in R^{2N}.$$

An allocation is called feasible if it satisfies the constraint (7.1). A feasible allocation $(c_1^*, a_1^*, \ldots, c_N^*, a_N^*)$ is Pareto efficient if there is no other feasible solution at which every country's utility is at least as high, and one's utility is strictly higher, than at $(c_1^*, a_1^*, \ldots, c_N^*, a_N^*)$.

A Pareto-efficient allocation must maximize a weighted sum of utility functions

$$W(c_1, \ldots, c_n, a) = \sum_{n=1}^{N} \lambda_n u_n(c_n, a)$$

with $\Sigma_n \lambda_n = 1$ subject to feasibility constraints. Varying the λ_n's, one traces out all possible Pareto-efficient allocations. The λ_n's are of course exogenously given welfare weights, and a standard set of weights is $\lambda_n = 1/N$ for all n. We are assuming in this formulation that utilities are comparable across countries. This means that we cannot change the units of measurement of utility in any country without making similar changes in other countries. Each country n faces a constraint in terms of allocating total endowments into either consumption, c_n, or atmospheric quality, a_n, represented by the function Φ_n. Then a Pareto-efficient allocation is described by a solution to the problem:

$$\max W(c_1, \ldots, c_n, a) = \sum_{n=1}^{N} \lambda_n u_n(c_n, a), \tag{7.2}$$

$$\text{subject to} \quad a_n = \Phi_n(c_n) \quad \text{and} \quad n = 1, \ldots, N \quad \text{and} \quad a = \sum_{n=1}^{N} a_n. \tag{7.3}$$

Note that, by definition, the marginal cost of abatement is the inverse of the marginal productivity of the function Φ_n:

$$MC_n(a_n) = -1/\Phi_n'(c_n). \tag{7.4}$$

A Pareto-efficient solution solves problem (7.2).

PROPOSITION 1 At a Pareto-efficient allocation $(c_1^*, a_1^*, \ldots, c_N^*, a_N^*)$, the marginal cost of abatement in each country, $MC_n(a_n^*)$, is inversely proportional to the marginal valuation of the private good c_n, $\lambda_n \partial u_n / \partial c_n$. In particular, the marginal costs will be equal across countries if and only if the marginal valuations of the private good are equal, that is, $\lambda_n \partial u_n / \partial c_n$ is independent of n.

PROOF. The solution to the maximization problem (7.2) must satisfy the first-order conditions:

$$\lambda_j \partial u_j / \partial c_j = -\left(\sum_{n=1}^{N} \lambda_n \partial u_n / \partial a \right) \Phi'_j$$

for each country $j = 1, \ldots, N$. Because at a Pareto-efficient allocation the expression $(\Sigma_{n=1}^{N} \lambda_n \partial u_n / \partial a)$ is the same constant for all countries, denoted K, and because, as noted in (7.4),

$$MC_n(a_n^*) = -1/\Phi'_n(c_n),$$

we have that a Pareto-efficient allocation is characterized by

$$MC_j(a_j^*) = \frac{K}{\lambda_j \partial u_j / \partial c_j},$$

and the proposition follows. ■

Proposition 1 shows that the product of the marginal valuation of private consumption and the marginal cost of abatement in terms of consumption is equal across countries. Writing this product $\lambda_j \partial u_j / \partial c_j \cdot \partial c_j / \partial a$, we see that it can be interpreted as the marginal cost of abatement in country j measured in utility terms, that is, in terms of its contribution to the social maximand $\Sigma_n \lambda_n u_n(c_n, a)$. An immediate implication is that in countries that place a high marginal valuation on consumption of the private good, typically low-income countries, the marginal cost of abatement at an efficient allocation will be lower than in other countries. If we assume an increasing marginal cost of abatement (diminishing returns to abatement), then this of course implies lower levels of abatements in poor countries than in rich countries.

Under what conditions can we recover the "conventional wisdom" that marginal abatement costs should be equalized across countries? We need to equate the terms $\lambda_n \partial u_n / \partial c_n$ across countries. This could be done by assumption: We can simply decide as a value judgment that is an input to the planning problem that consumption will be valued equally on the margin in all countries. Given the enormous discrepancies between the income levels in OECD countries and countries such as India and China and the need for all of them to be involved in an abatement program, such a value judgment seems most unattractive. It is, however, implicitly done in simulation models that seek to maximize world GNP or similar measures.

There is an alternative possibility. Modify the original problem to allow unrestricted transfers of private goods between countries:

$$\max W(c_1, c_2, \cdot\, c_n, \ldots, a) = \sum_n \lambda_n u_n(c_n, a)$$
$$\text{subject to } a_n = \Phi_n(y_n)$$
$$\text{and } a = \sum a_n \text{ and } \sum y_n = \sum c_n. \quad (7.5)$$

This is the same as before, except that we now distinguish between the consumption of the private good by country *n*, denoted c_n, and the production of the private good by country *n*, denoted y_n. These need not be equal. In addition, we now require the sum of the consumptions across countries to equal the sum of the productions $-\Sigma\, y_n = \Sigma\, c_n$, instead of having these equal on a country-by-country basis. By this modification we are allowing the transfer of goods between countries; that is, we are allowing lump-sum transfers. Note that this is not a model of international trade, which would require the imposition of balance-of-trade constraints. Clearly, the first-order conditions now are simply

$$\lambda_n \frac{\partial u_n}{\partial c_n} = v \forall n \quad (7.6)$$

$$\Phi_n' \sum \lambda_j \frac{\partial u_j}{\partial a} = -\, v \forall n \quad (7.7)$$

Set $K = \Sigma\, (\partial u_j / \partial a)$. Thus, from (7.6) and (7.7) we get

$$\lambda_n \frac{\partial u_n}{\partial c_n} = -\Phi_n' K \quad (7.8)$$

as before. However, we now have an extra condition (7.6), namely, $\lambda_n (\partial u_n / \partial c_n) = v\ \forall n$. Substituting this into (7.8) gives

$$v = -\Phi_n' K,$$

which of course implies that physical marginal cost is the same across all countries, as v and K are common to all countries. Thus, if we solve an optimization problem that allows unrestricted transfers between countries and make the transfers that are needed to solve this problem, it will then be efficient to equate marginal abatement costs.

Consider now the case of two countries, each with a Cobb-Douglas utility function,

$$u_n(c_n, a) = c_n^\alpha(a)^{1-\alpha} = c_n^\alpha(a_1 + a_2)^{1-\alpha},$$

where the abatement production function Φ_n is

$$a_n = \Phi_n(c_n) = k_n(Y_n - c_n)^{1/2}, \qquad k_n > 0, \text{ for } n = 1, 2,$$

for example, $k_1 = k$ and $k_2 = 1$. This allows us to accommodate potentially different efficiencies of abatement across countries. For simplicity the two countries are assumed to have the same utility function.

PROPOSITION 2 At a Pareto-efficient allocation, the fraction of income that each country allocates to carbon emission abatement must be proportional to that country's income level, and the constant of proportionality increases with the efficiency of the country's abatement technology.

PROOF. Our problem (7.2) can now be written as

$$\max_{c_1,c_2} W(c_1, c_2) =$$

$$\max\{c_1^\alpha[k(Y_1 - c_1)^{1/2} + (Y_2 - c_2)^{1/2}]^{1-\alpha} + c_2^\alpha[k(Y_1 - c_1)^{1/2} + (Y_2 - c_2)^{1/2}]^{1-\alpha}\}.$$

Let

$$A = [k(Y_1 - c_1)^{1/2} + (Y_2 - c_2)^{1/2}].$$

The first-order conditions for a maximum are then

$$\alpha c_1^{\alpha-1}A^{1-\alpha} - 1/2(Y_1 - c_1)^{-1/2}k\{c_1^\alpha A^{-\alpha}(1-\alpha) + c_2^\alpha(1-\alpha)A^{-\alpha}\} = 0$$

and

$$\alpha c_2^{\alpha-1}A^{1-\alpha} - 1/2(Y_2 - c_2)^{-1/2}\{c_1^\alpha A^{-\alpha}(1-\alpha) + c_2^\alpha(1-\alpha)A^{-\alpha}\} = 0,$$

which simplify to

$$\left(\frac{c_1}{c_2}\right)^{\alpha-1} = k\left(\frac{Y_1 - c_1}{Y_2 - c_2}\right)^{-1/2}.$$

Because $\alpha < 1$ this implies that for Pareto efficiency the income allocated to abatement by each country ($a_n = Y_n - c_n$, $n = 1, 2$) must be proportional to the income level, or the level of consumption, of the country (c_n). Furthermore, the larger is the abatement productivity of a country ($k = k_1$), the larger is its abatement allocation as a proportion of income. ■

7.3 Abatement Costs, Taxes, and Emission Permits

Although the atmosphere is a classic public good in terms of consumption, it is produced in a decentralized way, and the first-order conditions for efficient allocation and provision of this "good" are different from the classical ones and closer to those characteristic of a general externality, as modeled in Heal [7].

Once the optimal consumption/abatement levels in each country are found, then quotas on emissions could be assigned to each country on the basis of these levels, and permits could be issued and freely traded as financial instruments across countries on the basis of these quotas. A system of permits for carbon emissions has of course been contemplated for some time, but as far as we know the country-by-country quotas for these permits have not been connected to the optimality conditions for the allocation of public goods produced in a decentralized way. It would be desirable to ascertain what form of market organization for the permit market would be required to reach efficiency. For example, would it involve uniform pricing, as in a competitive market, or personalized prices, as in a Lindahl equilibrium? This should be a subject for further research.

References

1. Baumol, W. J., and W. Oates. *The Theory of Environmental Policy.* Cambridge: Cambridge University Press, 1977, 1988.
2. Chichilnisky, G. "North-South Trade and the Dynamics of Renewable Resources." *Structural Change and Economic Dynamics* 4, no. 2 (December 1993): 219–48.
3. Chichilnisky, G. "North-South Trade and the Global Environment." *American Economic Review* 84, no. 4 (September 1994): 851–74.
4. Chichilnisky, G. "The Abatement of Carbon Emissions in Industrial and Developing Countries." Paper presented at the International Conference on the Economics of Climate Change, OECD/IEA, Paris, June 14–16, 1993; see also *OECD: The Economics of Climate Change,* June 1993, referenced in chapter 1.

5. Chichilnisky, G., and G. M. Heal. "Global Environmental Risks." *Journal of Literature, Special Issue on the Environment* (fall 1993): 65–86.
6. Coppel, J. "Implementing a Global Abatement Policy: Some Selected Issues." Paper presented at the International Conference on the Economics of Climate Change, OECD/IEA, Paris, June 14–16, 1993.
7. Heal, G. "Economy and Climate: A Preliminary Framework for Microeconomic Analysis." *Agricultural Management and Economics: Commodity and Resource Policies in Agricultural Systems,* ed. R. Just and N. Bockstael. New York: Springer-Verlag, 1990.
8. Atkinson, A., and J. Stiglitz. *Lectures on Public Economics.* New York: McGraw-Hill, 1980.
9. Weyant, J. "Costs of Reducing Global Carbon Emissions: An Overview." *Journal of Economic Perspectives,* Symposium on Global Climate Change (1993).

Chapter 8

Differentiated or Uniform International Carbon Taxes: Theoretical Evidences and Procedural Constraints

Jean-Charles Hourcade
Laurent Gilotte

8.1 Introduction

From the late 1980s to 1996, debates on economic incentives aiming at curbing greenhouse gas emissions focused on a uniform international carbon tax. There are many historical reasons why attempts to coordinate climate policies through price signal failed and why coordination through quantitative emission limits was adopted at CPO3 (3rd Conference of the Parties, Kyoto 1997). The latter framework, however, is not firmly established as long as the following question is unresolved: which rules should be adopted for the distribution of primary rights to developing countries? If no politically acceptable rule can be found, the negotiation agenda may see the return of coordination through prices or some hybrid system. This paper aims at shedding light on the difficulties inherent to the price approach, some of which in fact are comparable with those impinging on quota-based coordination. Relying on a theoretical model that captures the key practical aspects of climate policies, this chapter demonstrates that an efficient allocation is achieved by differentiated taxes. Beyond existing uneven distribution of income, this is due to country-specific side effects of a carbon tax and specifics of development patterns. A uniform tax would be appropriate only if applied together with transfers between coun-

We thank Khalil Hélioui for discussions and comments on the subject of this paper.

tries. Considering the difficulty of negotiating such transfers, a uniform carbon tax would require each country be persuaded that this tax is welfare improving, thanks to the positive side-effects of removing existing distortionary taxes and to negative costs potentials. Beyond this short-term perspective, we point out the necessity of differential treatment (taxes on emissions from industry could be harmonized and taxes on households and transportation sector differentiated) to reconcile the objective of achieving equal marginal costs in welfare across nations and the necessity of nondistorting competition on international markets.

The insights provided by the toolbox of economists are obscured by the differences, often neglected in policy debates, between a first-best solution and solutions accounting for political constraints hinging on the negotiation process: sovereignty principle, subsidiarity principle (in the European Union), political judgments about the social acceptability of measures, loose and unstable perceptions of self-interests, and influence of intellectual traditions.[1]

The professional reflex of economists is to distinguish as clearly as possible these two levels of analysis, leaving to the policymaker the task of minimizing the gap between the first-best and second-best solutions. However, when this gap is too important, experience demonstrates that policymakers ultimately tend to disregard the results of economic analysis and to prioritize considerations of "procedural efficiency" such as the political acceptability and the simplicity of enforcement and monitoring of given policies.[2] To avoid this distrust in the case of climate policies, economists should consider the procedural constraints and transaction costs of specific policies at the outset of their analysis; however, symmetrically policymakers should also note that many counterintuitive conclusions of theoretical analysis helps us to understand why the expected procedural efficiency of some policy packages might not be realized.

The interest of this double requirement can be illustrated in the case of debates about internationally harmonized carbon taxes. Such a perspective was officially supported by the European Commission before the 1992 Earth Summit in Rio de Janeiro and was discussed further in the European Union after the Essen summit in December 1994.

The main rationale for this proposal is indeed procedural in nature and relies

---

[1]This is, for example, the case for the incentive systems: European countries accept seemingly the perspective of carbon taxes more easily than the United States and are reluctant to consider tradable emission permit systems

[2]This is one of the reasons why the overwhelming majority of environmental policies rely on so-called noneconomic instruments, such as regulation and standards, when economic literature advocates for economic instruments in the form of Pigouvian taxes or tradable pollution permits.

on a political intuition. Noting from economics that Pigovian taxes or tradable permit systems lead to the same optimum, the defenders of an international carbon tax call attention to the difficulties that an international tradable emission permits system (ITEPS) would confront: the capacity of the regulatory authority to impose sanctions, the disagreements about the currency to be traded (only carbon dioxide [CO_2] or all greenhouse gas sources and sinks counted in terms of CO_2 equivalent), or the risks of monopolization of the market by oil producers. However, the Gordian knot of the system is obviously the agreement on the initial allocation of permits. First, each of the possible criteria for this allocation (grandfathering, egalitarian, and a two-tiered approach)[3] might be unacceptable to a significant number of negotiating parties. Second, decisions on the initial allocation and on the definition of the traded currency are not independent of each other. The scope of the system and the global warming potentials of different gases used to aggregate emissions change countries that would be net payers and net receivers in an ITEPS.[4]

A uniform carbon tax is meant to provide a clear economic signal that would not distort international industrial competition and that could be implemented without excessive administrative costs. Moreover, despite the fact that a significant part of the literature examines tax systems whose product is internationally redistributed, political constraints on the acceptability of the system (e.g., sovereignty principle, reluctance to accept such transfers in the name of very long run issues, and monitoring the use of funds) explain why all the official proposals to date assume that these taxes should be internationally coordinated but that their revenue would be internally recycled in each country.

Our purpose here is not to refute the interest of uniform international carbon taxes but to show why its procedural efficiency is not so evident as it seems intuitively. It could indeed be deeply undermined by the equity-efficiency dilemma even if it does not confront it as directly as in the case of property rights assignment on atmosphere.

The basic reason is the well-established result that it is not easy in practice to separate efficiency from distribution when goods are public. This is the framework for climate policies simply because of the physical fact that GHG atmospheric concentration is the same for all of us independently of our level of income and our level of concern for climate change. The policy implications of this point for income distribution have been to date developed by Chichilnisky [3] and Chichilnisky and Heal [4].

---

[3]On this topic, see OECD [21].

[4]This was illustrated by the controversy between Anyl Agarwal and the World Resources Institute about the role of the CH_4 and deforestation in the ranking of GHGs emitters.

We develop a theoretical framework capturing other reasons why uniform carbon taxes are, in the general case, suboptimal if not accompanied by lump-sum transfers among countries. We focus on the concept of abatement technology and on the difficulties stemming from the concept of "double dividend" yielded by the recycling of the revenues of a carbon tax; in a further step, coming back to procedural efficiency, we point out some paradoxes likely to be involved in the international negotiation of a carbon tax.

8.2 Climate Policies and Limits of First-best Framework

Beyond the appeal of "procedural efficiency," a second reason that the focus has been so easily placed on a uniform tax is that this solution corresponds to a widely held view among environmental economists. In a Pigovian perspective a tax gives to every agent the same "signal" about the potential costs of climate change. Conventionally, it is then assumed to allow for an optimal allocation of abatement efforts because agents will adopt only the GHG abatement techniques whose marginal cost is lower than the tax level. However, this allocative efficiency of a uniform carbon tax can be questioned because of the specifics of the climate change issue.

8.2.1 Climate as a Public Good: Theoretical Backgrounds of a Recent Controversy — A first criticism of tax uniformity stems from the heterogeneity of existing fiscal systems. From a basic demonstration relying on the Harberger triangle, it can be shown that a uniform tax would place a bigger burden on those countries whose preexisting energy taxation levels are high (Hoeller and Coppel [12]).

A formal solution to this problem has been sought in public finance theory: When a tax aims at internalizing an externality and at levying funds for government's budget, the optimal tax structure should be additive. The externality-creating commodity should be subjected to a tax which is a weighted average of two terms—one "fiscal" and the other equal to the marginal social damage, as in Pigovian taxation (Sandmo [21]).[5] "Therefore, there is no reason to try to achieve equality of the total fiscal burden on fossil fuels in different countries. On the other hand, there are grounds to seek agreement on the amount of the reference internalizing tax" (Coppel [5]). A uniform carbon tax could be simply added to fiscal systems previously restructured according to Sandmo's rule.

[5] A summary of this discussion can be found in Godard [9].

Even without considering the implementation difficulties and transaction costs associated with this solution, its fundamental caveat comes from the fact that Sandmo's framework assumes that individuals face the same prices and taxes. In contrast to this, the point made by Hoeller and Coppel [12] derives directly from the fact that countries have different consumer prices, partly as a consequence of their individual fiscal systems. With another approach, emphasizing the consequences of differences in income levels (with subsequent heterogeneous preferences) and of uneven access to technology, Chichilnisky [3] demonstrated that, in the case of climate policies, a uniform internalizing tax would lead to a suboptimal equilibrium if there are no lump-sum transfers among countries.

Chichilnisky's argument surprised many economists; Bohm [2] recalled that trade among nations allows for optimality of uniform taxation despite differences in utilities. The rationale for such a solution is the following: If international markets are assumed to ultimately give each country access to the same technology basket (either directly or by assuming a free access to the goods produced by the best available techniques), the technology mix apt to minimize the overall cost of meeting a given abatement target will be implemented only if the same price signal, equal to the marginal abatement cost, is given to each agent. This view prioritizes the launching of a clear signal so as to optimize the technology mix. Nevertheless, Bohm's model implicitly resorts to lump-sum transfers so as to equalize the marginal utility of income among countries. This result is confirmed by Chichilnisky and Heal [4] who, in a model allowing for the transfers of goods between countries (lump-sum transfers), state that "if we make the transfers that are needed to solve this problem, it will be then efficient to equate marginal abatement costs" (i.e., to equate carbon taxes). They also point out (p. 447) that a model allowing lump-sum transfers "is not a model of international trade, which would require the imposition of balance of trade constraints."

In the specific case of climate policies, it can be argued that a uniform tax would be an optimal solution only under very exceptional conditions. The systematic demonstration can be derived from a very general model with a private good generating externalities written by Laffont [18, pp, 75ff.]. Interestingly for the current discussion, he points out a special type of externality that could encompass climate change issues: In the case of "nonpersonal externalities," it can be shown that first-best (Bowen-Lindahl-Samuelson conditions) is achieved through a uniform Pigovian tax and lump-sum transfers (the net amount of the transfers being equal to the total revenue of the tax). Each Pareto-optimal level of emission is jointly determined with a uniform tax and a set of lump-sum transfers.

In the case of a compensation mechanism accompanying an international carbon tax, there is little chance that countries will share their tax receipts so as to achieve an allocation of income that enables an optimal emission level corresponding to the tax. This prevents a uniform taxation from being automatically optimal. There is indeed no reason that the necessary transfers will be acceptable, and thus that optimal emission level will be achieved. Intuition suggests on the contrary that this is true only for abatement targets and corresponding tax which do not imply too-large transfers. For a given distribution of income and in a perfect information context, any first-best solution that leads to welfare gains for every country is potentially acceptable, whatever the necessary initial transfers. But in practice, because of the informational and procedural constraints, the scope for negotiable first-best solutions is much narrower and perhaps void.

This doubt about the likelihood of international transfers restoring systematically the optimality of a uniform carbon tax is obviously strengthened if one considers the implementation difficulties and transaction costs of such a solution. This leads us to investigate further the meaning and the relevance of the heterogeneity hypothesis about both preferences and production functions of the GHG emissions abatement that underpin the plea for differentiated taxations.

We do not come back in the rest of this text to the inequality of the wealth distribution as a sufficient reason for differentiated taxes if appropriate compensations are not given to offset the recessive impacts of a uniform tax. This result is well established by Chichilnisky and Heal, but the level of abstraction of the abatement production function of their model might obscure the fact that, if one considers seriously the determinants of emission trends, their line of argument stands even in a world with equal income distribution levels and a free trade ensuring equal access to the best available abatement techniques and to the composite goods produced at the lowest cost. The crux of the matter is the linkages between the content of the production functions of GHG abatement and the reasons for heterogeneous preferences for energy.

8.2.2 Specifics of Climate Issues and Policies — Even in a perfect world market economy, the transformation frontier between GHG abatement and other goods and services never is the same for all countries. The first-best solution by which the best available techniques can be implemented by each country or by which, in the case of nontransferable techniques,[6] international

[6]Natural comparative advantages resulting, for example, from the specifics of the geographic context, such as the endowments in hydropotentials.

trade gives each country access to the goods at the same price (equal to its marginal production costs) comes to a centrally planned abatement program mobilizing a unique set of techniques ranked by decreasing returns. The existence of barriers in technology markets would not be, per se, an argument against a uniform tax but a reason to take up transitory measures aimed at removing market imperfections.[7]

Note that, in this framing, the production frontier governing the transformation rate between GHG abatement and other goods and services results strictly from a given set of techniques defined in a pure engineer's sense; the abatement costs curve is calculated as the arithmetic sum of technical costs. This framing stops being relevant when one accounts for two specifics of the debates about greenhouse policies: (1) In the energy field, the very definition of an abatement technique is less trivial than it seems because of the fact that the ranking of technical solutions by decreasing cost-effectiveness is very conditional on assessing the cost of delivering a given physical quantity of final energy or assessing the cost of providing a given set of end-use services, and (2) there are critical debates about possible economic double dividends (or extra macroeconomic costs) of climate policies and about the magnitude of a wedge between their gross and net costs.[8]

Heterogeneity of Utilities and Abatement Costs

As professional economists, our first reflex is to frame a public policy problem in a way that separates agents' utilities on the one hand and technical abatement costs on the other. However, in the case of the greenhouse issue, this separation is not as easy as it seems at first glance, for reasons that are easy to illuminate in the case of energy systems.

From the mid-1970s on, "bottom-up" specialists in the energy field helped us understand that substitution elasticities between other goods and final energy described by current statistics might be a misleading indicator of the driving forces behind energy demand. This argument was basically used to point out efficiency gaps along the transformation chain between primary energy, final energy, and end-use energy services. The heated energy policy debates with "top-down" specialists about the meaning and magnitude of these effi-

[7]This view neglects possible obstacles to technology appropriation pointed out by literature on appropriate technology; these are not, stricto sensu, due to imperfections in international markets but to parameters, such as prevailing institutions, technical capabilities, and cultural habits, which make the hidden costs of using a given technology different in various countries.

[8]For a taxonomy of costs concepts in use in debates about climate policies, see chapter 8 of working group III of the Ipcc report (Hourcade, Richels, Robinson 1995).

ciency gaps[9] masked a very important theoretical implication of this move in energy demand analysis. It is uncontestable that, theoretically, neither final energy demand nor energy services should be included in the consumer's utility function. They are ancillary services of components of this function, such as transportation, thermic comfort, or food conservation. In this sense there is never a substitution between energy and other goods in the individual preference function (e.g., between a fried egg and the energy needed to cook it) but between energy-intensive and non-energy-intensive products and services and between energy-intensive and non-energy-intensive ways of making them.

This forces us to question the definition of GHG abatement techniques and of revealed preferences in models, including energy in the welfare function, and paves the way for an argument in favor of differentiated taxes even in a world without income inequalities. In current energy demand functions, the apparent substitution elasticity between energy and other goods should not be interpreted as a measurement of the "pure preference" for energy. In the case of a high energy price increase, for example, a medium town with a tramway network could switch rather quickly to less energy intensive transport systems (combining tramway and bicycle) when a town typically built for cars, such as Los Angeles, would simply not be able to do so over the same time period. Even if the citizens of the two cities had the same degree of concern for climate change, the revealed willingness to substitute non-energy-intensive goods and services to gasoline will simply be higher in the first one.

If we stand within this framework (substitution between energy and a composite good), it follows that the observed preference functions differ across countries for reasons other than differences in income and "pure" preference for precaution toward climate risks. A more appropriate theoretical framework would obviously be to treat urban structures and transportation modes as technical endowments. The Appendix gives a very tentative formalization of this issue with a world composed of agents with identical utility functions, including a composite good, leisure activities, and thermal conditioning (heating or cooling), but living in national contexts whose features (manmade or natural features do not matter at this stage of analysis) demand different quantities of energy and transportation for achieving the same level of welfare.

The critical policy implication comes from the fact that, contrary to energy-efficient technologies that can be adopted at the margin of a system, urban and transportation systems constitute technical systems in the Gille [8] sense with their internal systemic coherence. The perfect international market hypothesis

[9] A very useful clarification of this debate can be found in Jaffe and Stavins [16].

is then inapt for solving the problem of the heterogeneity of abatement costs because these "technologies" are not importable goods: Los Angeles cannot "import" an "urban structure technology" even over the middle term. Then, because abatement costs do not depend only on technical answers in the engineering sense (costs of switching from coal to gas or nuclear in electricity generation or costs of more efficient boilers), the production function of GHG abatement cannot be homogeneous and is inherently country specific.

Double-Dividend of Tax Recycling and Macroeconomic Production Function of GHG Abatement

For defining the content of the production function of GHG abatement, another source of complication is the fact that, given the amount of uncertainties surrounding climate change, policy debates were underpinned by the search for so-called no-regret policies, namely, for policies entailing no net incremental cost and that will not be regretted if, ultimately, anthropogenic climate change is proved to be harmless.

The no-regret concept results from a pure strategic intuition and has no rigorous definition. It will suffice, for the following discussion, to note that if the current state of economy is assumed to be optimal, an improvement of environmental quality is possible only through a reduction of production of conventional goods. A no-regret climate strategy is possible only if this economy is located somewhere below the theoretical production frontier describing the maximum of production of conventional goods for a given quality of environment and if the policy choice enables the progress toward the production frontier in order to reduce GHG emissions.[10]

Initially centered on the "efficiency gap" and possible negative costs measures, discussions about "no-regret" were extended to the environmental double dividend expected from the side effect of GHG reductions on other environmental issues (e.g., acid rain, tropospheric ozone, and urban congestion) and, with more heated disputes, from the economic double dividend of recycling the revenue of carbon taxes.

A tax on CO_2 emissions is indeed meant to be an incentive to foster the use of carbon saving technics and not a financial source for supporting research on energy efficiency or supplying a world fund, such as the Global Environment Facility. This tax should be high enough to have an effect on consumption and

[10] An overview of this debate can be found in chapters 8 and 9 of the forthcoming IPCC report (Hourcade, Richels, and Robinson [15]).

production choices, and this poses critically the question of how its revenue is recycled.

The side effects of an internally recycled ecotax[11] were analyzed in great detail by some empirical macroeconomic studies, mainly in the European context, that concluded positively about a double dividend. This was the case in the Quest simulations by the European Commission as well as in national studies in countries such as Germany (Walz et al. [22]), France (Godard and Beaumais [10]), and the United Kingdom (Barker [1]). Works from a theoretical perspective shed some doubts on the likelihood of such a double dividend being apt to offset the gross costs of climate policies if all the general equilibrium effects of such a fiscal reform are accounted for. This is not the place to enter into the details of this discussion but it is noncontroversial that a double dividend occurs when the marginal distortionary effect of a carbon tax (or ecotax) is lower than the distortionary effect of taxes for which it is substituted.

This introduces a second element of heterogeneity between countries' cost functions: Tax impacts are the net result of the costs of increasing energy prices and of the benefit from removing more onerous taxes, and both these parameters are mostly country specific. Many European countries, for example, finance not only their public administration but also their health system, social security, and teaching system by raising funds from taxes levied directly or indirectly on wages; this wedge between the labor cost and the net wages might be a cause of structural unemployment. The fiscal system is very different in the United States and Japan as a practical translation of different views of social organization.[12] In the same way, the measurement of the distortionary effects of preexisting energy taxes cannot be directly derived from their observed level, as many oil-importing countries levy energy taxes to achieve public objectives, such as security, minimization of shocks of trade balance, and funding of road infrastructure.

What matters here is that the direct costs of abatement are not the only costs a government must face. The recycling of a carbon tax creates a wedge between the gross cost of GHG abatement (the sum of the costs of abatement technology) and the net cost for the economy; determinants of this wedge are country specific and are not apt to be homogenized through foreign trade[13] across countries because the double dividends are intangible.

---

[11] Note that ecotaxes other than carbon tax have been studied (e.g., the carbon energy tax in Europe).

[12] On the difference between U.S. and European contexts, see Krugman [17].

[13] These components characterize a second-best world and would not play a role any more if fiscal distortions were removed in all countries prior to climate policies and if the rules for interpreting preexisting energy taxes were the same in all countries. These preconditions will be hardly fulfilled prior to the forthcoming negotiation steps.

8.3 Two Tentative Models

These features of the international negotiation over a carbon tax can be illustrated with two versions of the same generic model.

8.3.1 A One-Energy Model— Let us assume an economy with two goods available: Q, the composite good, and F, fossil energy. Each country is represented with (1) a utility function $U_i = U_i(Q_i; F_i; G)$, G being a public externality, or the GHG emissions, with $UG < 0$, and (2) an income level R_i.

The planner aims at choosing a tax set (t_i) on fossil fuel consumptions (F_i) that maximizes the aggregate welfare function. Because of the procedural constraints stated earlier, he is not allowed to make lump-sum transfers between countries. But he gives each country a compensation as large as the tax revenue he collects in the country. The planner announces to the countries the individual taxes (t_i) and the level of externality G they will face, as well as the level of compensations (C_i) they will receive.

Each country maximizes its utility choosing its demands for Q and F. It does not account for the externality it produces itself. The budget constraints are

$$R_i \geq Q_i + \pi_F(t_i)F_i \quad \text{with } \pi_F = (p_F + t_i) \quad \text{and} \quad R_i = R_i^o + C_i.$$

We write the demand functions in Q and F as

$$q_i(\pi_F(t_i); R_i; G) \quad \text{and} \quad f_i(\pi_F(t_i); R_i; G)$$

and the indirect utility functions as

$$V_i(\pi_F(t_i); R_i; G).$$

The planner must satisfy two constraints. First, the emission level (by approximation the sum of fossil fuels consumption), resulting from the individual countries' optimization, should not exceed the level chosen by the planner (denoted G). Second, the compensation (denoted C_i) granted to a country should not exceed the tax receipts in this country. His maximization program is

$$\max_{t;C;G} \sum_{i=1}^{n} \alpha_i \cdot V_i(\pi_F(t_i); R_i; G)$$

with constraints

$$\sum_{i=1}^{n} f_i(\pi_F(t_i); R_i; G) - G \leq 0 \quad \text{and} \quad C_i - t_i \cdot f_i(\pi_F(t_i); R_i; G) \leq 0.$$

Calling μ and v the Lagrange multipliers associated with the previous constraints, the first-order conditions for a t; C; G solution of the maximization problem are

$$\forall i \qquad 0 = \alpha_i \frac{\partial V^i}{\partial \pi_F} - \mu \frac{\partial f^i}{\partial \pi_F} - v^i\left(-f^i - t^i \frac{\partial f^i}{\partial \pi_F}\right), \tag{8.1}$$

$$\forall i \qquad 0 = \alpha_i \frac{\partial V^i}{\partial R} - \mu \frac{\partial f^i}{\partial R} - v^i\left(1 - t^i \frac{\partial f^i}{\partial R}\right), \tag{8.2}$$

and

$$0 = \sum_{i=1}^{n} \alpha_i \frac{\partial V^i}{\partial G} - \mu\left(\sum_{i=1}^{n} \frac{\partial f^i}{\partial G} - 1\right) + \sum_{i=1}^{n} v^i t^i \frac{\partial f^i}{\partial G}. \tag{8.3}$$

Multiplying the second equation by f_i and adding it to the first and then applying Roy's lemma

$$\frac{\dfrac{\partial V_i}{\partial \pi_F}}{\dfrac{\partial V^i}{\partial R}} = -f_i,$$

we obtain

$$\forall i \qquad 0 = (-\mu + v^i t^i) \cdot \left(\frac{\partial f^i}{\partial \pi_F} + f^i \cdot \frac{\partial f^i}{\partial R}\right),$$

that is,

$$\forall i, \qquad t^i = \frac{\mu}{v_i} \quad \text{with} \quad v_i = \alpha_i \frac{\partial V^i}{\partial R} = \alpha_i U^i_Q \quad \text{and}$$

$$\mu = -\sum_{i=1}^{n} \alpha_j \frac{\partial V^j}{\partial G} = -\sum_{i=1}^{n} \alpha_j U^j_G,$$

and then

$$t^i = \frac{-\Sigma_{j=1}^{n} \alpha^j U^j_G}{\alpha_i U^i_Q} \quad \text{and} \quad \forall (i, j), \ \frac{t^i}{t^j} = \frac{\alpha_i}{\alpha^j} \frac{U^j_Q}{U^i_Q}.$$

This result is identical to the result of Chichilnisky and Heal ([4], p. 446); the taxes are differentiated, and, comparing two countries, their tax ratio is equal to the inverse of the ratio of their marginal utility of numeraire. Thus, a tax is equal to the ratio of social benefit of a marginal abatement to marginal utility of income. Obviously, if one admits that the tax revenue can be redistributed (and not only returned), then, for all $i \in N$, $v_i = v$, and the tax is uniform.

Note that these taxes, which are equal to the marginal cost of abatement in each country, should not be confounded with the marginal social cost of abatement, which is the same for every country (as being equal to the multiplier μ) and is equal at the optimum to marginal social benefit from this abatement. Keeping pollution and taxes constant, the marginal welfare cost of abatement for country i is $\alpha^i(U_F + t_i U_Q - \pi_F(t_i)U_Q)$, and reducing consumption in F leads to lower tax receipts (and thus the received compensation) but to higher available income to consume more numeraire. Each country maximizing its utility, we have $U_F - \pi_F(t_i)U_Q = 0$, so the cost is

$$\alpha^i \frac{dV^i}{df}\Big|_{t;G} = (\alpha^i U_Q^i)t^i = -\sum_j \alpha^j U_G^j.$$

A tax can also be written as the ratio of marginal disability of abatement to marginal utility of income:

$$t^i = \frac{\dfrac{dV^i}{df}\Big|_{t;G}}{\dfrac{\partial V^i}{\partial R}}$$

8.3.2 Accounting for Side Effects — In the following version of the model, we add macroeconomic effects of the taxes: loss of competitiveness on international markets, sectoral adaptation, and double dividend from the recycling of a carbon tax. These net macroeconomic impacts can be interpreted as a reduction or an increase in the income available for consumption: $R_i(t_i) = R_i(0) - S_i(t_i)$, where $S(t)$ are the side effects associated with the tax ($S' > 0$).

The previous first-order condition (8.2) becomes

$$\forall\, i \in [1, n], \qquad 0 = \alpha_i\left(\frac{\partial V^i}{\partial \pi_F} - S_i' \frac{\partial V^i}{\partial R}\right) - \mu\left(\frac{\partial f^i}{\partial \pi_F} - S_i' \frac{\partial f^i}{\partial R}\right) - v^i\left(-f^i - t^i\left(\frac{\partial f^i}{\partial \pi_F} - S_i' \frac{\partial f^i}{\partial R}\right)\right); \qquad (8.4)$$

added with (8.2) and multiplied by $(f_i + S'_i)$, it gives

$$\forall i \in [1, n], \qquad t^i = \frac{\mu}{v_i} + \frac{S'_i}{\left(\dfrac{\partial f^i}{\partial \pi_F} + f'_i \dfrac{\partial f^i}{\partial R^i}\right)}$$

and thus $t^i \leq \mu/v_i$ because the denominator of the additional term is negative and S' positive in the general case.

Once again the taxes are differentiated and the marginal social costs of abatement are equal. When taxation generates side effects, the planner should levy a tax all the less high, as these costs are important at the margin (see the numerator of the additional term) and the marginal impact of the tax on consumption of F is limited (see the denominator, equal to the compensated variation of energy consumption). This confirms the nonoptimality of a uniform tax when countries are distinguished by their systems of preferences or macroeconomic reactions to the tax.

8.4 Policy Implications: Some Paradoxes about Negotiability

It follows that, if tax revenues are assumed not to be redistributed internationally, the optimal character of a uniform tax is challenged by many factors of heterogeneity: differences in marginal utility of income and in marginal utility of energy services, uneven access to the best available technologies, and country-specific side effects (additional cost or benefit) of a tax. These factors are apt to pose a problem of procedural efficiency of negotiating a uniform carbon tax if one accounts for the following constraints on policymakers:

First, in terms of aggregated welfare, collective optimum is reached when the marginal welfare cost of abating is equal across countries. As a consequence, countries characterized by a low price elasticity on the demand side or a low technical flexibility on the supply side should be conceded lower taxes (or higher compensations) whereas most of the emission abatement should be operated by countries characterized by high energy price elasticities and high substitution potential between fossil energies and carbon-free energies.

Second, this would not raise any problem if information about price elasticities and technical potentials could be easily revealed. However, parameters determining the balance of gains and losses for each country are far from being tangible. First, energy economists know that long-run income and price elasticities are very controversial issues, especially the respective weight of "autonomous" technical progress and "price-induced" technical progress, and, second, the double dividend of the recycling of a carbon tax, which is macroeconomic in nature, is by definition not observable ex ante.

Third, in a context in which each government must overcome domestic conflicts to adhere to an international agreement, even in the case of a no-regret policy,[14] there might be an asymmetry between the convincing power of intangible parameters surrounded by hard controversies and the symbolic value attached to the tax level. This level is likely to be viewed as an indicator of the required effort and of the risks in terms of international competitiveness; to put it another way, the tax level is tangible, and the economic double dividend will remain both intangible and controversial to the "losers" of such a change in the fiscal system.

Finally, governments will then tend to adopt strategic behavior so as to maximize compensation that they will receive; they will put to the forefront all kinds of arguments demonstrating the low elasticity and low technical adaptability of their country. This would undermine a negotiation process both on a differentiated tax and on compensations accompanying a uniform tax.

Thus, there is no obvious reason that such compensations will be more easily negotiated than differentiated taxes. A system of country-specific taxes or a uniform taxation accompanied by transfers across countries are formally equivalent in terms of quantity and quality of the required information. Deriving absolute conclusions about relative procedural advantages and deadlocks of each system is beyond the scope of economic analysis and would imply considering the sociological, institutional, and cultural determinants of the acceptability of each system.

We can nevertheless make some steps forward. After the Conference of Parties decisions in Berlin (March 95), we are indeed engaged in a sequential process. We must decide today not what will be the optimal solution for the twenty-first century but rather the first step of a precautionary strategy that allows for further adaptations and corrections. This is supported by theoretical research (Manne and Richels [19]; Hourcade and Chapuis [14]) and suggests that the search for strict no-regret policies should prevail over the short run.[15]

If no international transfers are operated, the adoption of no-regret measures

[14] A no-regret policy, which is supposed to remove current market and institutional imperfections to improve environmental quality without decreasing the size of the economy, is not a "free lunch": It implies paying the transaction costs of removing these imperfections, which might be politically sensitive in many circumstances.

[15] This is not in contradiction with the conclusion of the forthcoming report by The Intergovernmental Panel on Climate Change (IPCC) that we have now sufficient scientific understanding of the risks associated with global warming to plea for actions "beyond no-regret." What matters here is (1) that launching no-regret actions now is the maximum that can be expected given the actual degree of concern of international community toward climate change and (2) that implementing these policies now while triggering R&D on low-carbon-emitting technologies does not affect our capacity to mitigate climate change through more drastic and costly GHGs abatements in a second step.

in each country is strictly Pareto improving and are paradoxically restricted to policies that should be adopted regardless of any international coordination. In practice, however, such coordination is necessary for two reasons: (1) Possible distortions might occur in world markets at sectoral levels, and (2) the adoption of a no-regret policy, although yielding net collective benefits, entails transaction costs that a government will more easily cope with under the pressure of an international process.

At this stage the most sensitive distributional issues can be neglected and a learning process triggered along with two dimensions: time progressiveness and geographical progressiveness. Time progressiveness of the taxation is required to minimize adaptation costs and to enable countries to experiment with respect to the outcome of a fiscal reform; geographical progressiveness is the process by which an initial coalition of concerned countries demonstrating the effectiveness of the no-regret policies could be increasingly expanded.[16]

A uniform tax system might be preferred for avoiding distortions in international competition on energy-intensive industries; however, to be actually implemented, such a system will have to fulfill two conditions: very low compensation transfers and low expected net macroeconomic costs. The risk is an agreement on the lowest common denominator. The framework adopted by the European Union at Essen opens an alternative pathway: A differentiated tax system leaves each country to judge what tax level is compatible with a no-regret policy, and the most concerned countries might choose high tax levels, whereas others might choose low ones, lessons from experience being progressively derived by each of them. However, because of the risks of sectoral distortions, such a system is apt to lead to significant tax levels only if differences between these levels are not too high during too long time periods; otherwise, internal pressures will incite governments to lower their initial commitment.

It is only in a second step that the most sensitive issues raised by climate policies cannot be avoided by resorting to the no-regret concept. It is reasonable to expect that most of the no-regret potentials will be exhausted. The only one remaining factors of heterogeneity will be uneven income distributions and differences in utility of energy services stemming from differences in development patterns and lifestyles.

This might be the source of a very sensitive controversy not only between

[16]Numerous analysts think that the emergence of an anticarbon coalition will remain the most likely outcome of the current process. This first coalition, for example, a part of OECD countries, would try to expand step by step to other countries by bilateral negotiations; these would finally result in differentiated implicit prices for carbon; Coppel [5] imagines, for example, a G7 coalition negotiating with Russia, India, and China; these 10 countries are responsible for the three-quarters of the world carbon emissions from energy.

developed and developing countries but also among OECD countries.[17] The underlying critical question is whether heterogeneous preference levels for energy (and other GHG-emitting commodities) stem from objective natural conditions (rigorous climate and low human density) or from manmade infrastructures (geographical distribution of human settlements) or cultural habits (preference for high-powered cars). It is hardly questionable that infrastructure decisions in urban planning, transportation, telecommunications, and energy systems historically lead to contrasted energy consumption levels between countries having rather similar development levels (e.g., between North American and western European countries). An acceptance of differentiated taxes by low-emitting countries for reasons other than natural parameters is then unlikely because this would come to give up correcting structural determinants of energy consumption and to recognize a status of intangibility to habits and behavior of countries making a more profligate use of energy and to give them high-emission rights. Symmetrically, because of the welfare costs entailed during a very long transition period, policies that lead to an increasingly uniform carbon tax will be accepted by high-emitting countries only under the proviso of compensations. The difficulty is that the magnitude of this compensation is likely to be so high that it might not be accepted by other countries and generate centrifugal forces paralyzing the negotiation.[18]

---

[17]European expertise accepts, for example, more easily the perspective of reduced (or stabilized) energy consumption than the United States does while starting from a far lower benchmark level.

[18]In an illustrative exercise carried out in 1992, we calculated the global cost for switching from the World Energy Council (WEC) 1986 projections baseline scenario to the normative consumption target proposed by bottom-up modelers for the end of the twentieth century, Dessus and Pharabod [7]. We assumed that the additional efficiency progress (beyond the autonomous progress) would be triggered by energy taxes. In a first simulation we translated the WEC assumptions in a simple formalized expression of the respective role played by prices and autonomous factors in the energy-economy growth decoupling. The results reflected quite well the conventional wisdom prevailing in each region. The United States, Europe, and Japan would keep being very unequal consumers of energy per capita for similar levels of income (no homogenization of consumption patterns); responses to prices would be high in western Europe but weak in the United States; and autonomous technical progress would be high in Japan. In a second simulation we assumed that the view of each country on its own future could be contested and that responses to prices should be higher as the initial per capita energy consumptions are high and to decrease up to an asymptotic value of zero when price increase. The computation of this process toward homogenization of consumption patterns is based on endogenous price elasticities as a function decreasing with the achieved energy efficiency level.

In both cases differentiated taxes lower drastically the total costs of meeting the target at the world level compared to the cost of a uniform tax, but the distributive effects are totally contrasted. However, under assumptions WEC a uniform tax puts the brunt of the total cost on the United States whereas the implementation of differentiated taxes leaves it exempted in 2020 and multiplies its reference energy expenses by a factor of 2.2 only in 2060 to compare with a factor 3.4 for Europe and 3.2 for Japan in 2060. Under the assumption of long-term converging behaviors, the energy expenses are multiplied in 2060 by a factor 2.3 for the United States, but only 1.5 for Europe and 1.2 for Japan when implementing differentiated taxes.

Theoretical analysis suggests only one way out. It is a matter of fact that the main argument against differentiated carbon taxes is the risk of distorting international competition. By chance, making a distinction between energy demand from industry (roughly internationally tradable goods) and energy demand from households and transportation (roughly noninternationally tradable services), the results of the previous sections justify the use of harmonized taxes for industry and resorting to differentiated taxes for final consumers.

If one admits indeed that international markets progressively ensure the harmonization of techniques, then there is a collective interest in avoiding distorting competition (we are close to the first-best world with a total flexibility in trading and technical choices). This is not the case for households and transportation. Then the adoption of differentiated taxation as a function of the level of income and current development patterns is appropriate. However, the progressive convergence of consumption patterns will remain the stumbling block for the process, and the contradictions of interest among nations will be reduced only under the assumption that a high-and-fast technical change induced by price signal [19] is both equitable (the citizen of a rich country will pay more for climate mitigation than the one of a poor country) and efficient in terms of welfare costs.

It is worth mentioning that this conclusion is centered on an issue that is also present in today's discussions around a quota-based approach. Progressive convergence of per capita allocations also implies long-term convergence of energy consumption behaviors. For emissions quota as for differentiated taxes, organizing progressive convergence might be a realistic pathway if, simultaneously, a low-carbon-intensive technical change induced by climate policies grows fast enough to narrow the costs of abating GHG and the contradictions of interests across nations.

Appendix
Revealed Preferences for Energy Services and Development Patterns: A Tentative Model

We assume two countries that have the same utility function $U_i = U_i(Q_i; L_i; H_i)$ and consume a composite good Q_i, leisure activities L_i, and thermal well-being (heating or cooling). To be fulfilled, leisure requires a certain quantity of energy (of transport), $E = \Psi(L, \sigma)$, depending on the shape σ of the city (how

---

[19]The induced technical change means that the outcome of a steady price signal will move along a given production function and generate a new production function. The importance of this biased technical change on the costs of climate policies was pointed out by Hourcade [13] for France and by Goulder and Schneider [11] for the United States.

diffuse is the city). Heating/cooling requires also a certain quantity of energy depending on the harshness of the local climate conditions φ: $E = \varsigma(H, \varphi)$. Thus, the total energy consumption of country i is

$$E_i = \varphi(L_i, \sigma_i) + \varsigma(H_i, \varphi_i).$$

We assume that $\varsigma_H > 0, \varsigma_\varphi > 0, \varsigma_{H\varphi} > 0, \varphi_L > 0, \varphi_\sigma > 0, \varphi_{L\sigma} >$. Each country maximizes utility under the budget constraint $R_i \geq p_Q Q_i + p_E E_i$ with λ the associated Lagrange multiplier. The first-order conditions are:

$$U_Q = \lambda p_Q,$$

$$U_L = \lambda \varphi_L p_E,$$

and

$$U_H = \lambda \varsigma_H p_E. \tag{A8.1}$$

As the parameters σ and φ appear in the apparent prices of the needs, we can be sure that it will usually result in different levels of consumption across countries.

The planner aims to choose a tax set (t_i) on fossil fuel consumptions (E_i), limiting the emissions to a chosen level. He must obtain a minimal agreement from the involved countries and so does not allow to proceed in lump-sum transfer between countries, but having collected the taxes, he returns to each the revenue of its own tax.

Considering its endowment, the prices and the tax, and the compensation announced by the planner, each country maximizes its utility choosing its demands for Q and E.

Its new disposable income is now $R_i = D_i + C_i$, where D_i is the initial endowment and C the received compensation for tax, and the new price for E: $\pi_E = p_E + t_i$.

The planner has to keep in with two constraints. First, the actual emissions level, which results from the individual countries optimisation, should equal the global level chosen by the planner (we suppose hereafter that the emissions are proportional to the energy consumption). Second, the compensation granted to a country should equal the tax receipts in this country. Letting E^* be the energy/emissions goal, the maximization program of the planner is

$$\max_{t;C} \sum \alpha_i V_i(p + t_i; R_i) \quad \text{with constraints}$$

$$\sum E_i(l_i; h_i) - E^* \leq 0 \quad \text{and} \quad \forall i \; C_i - t_i E_i(l_i; h_i) \leq 0.$$

Calling μ and ν the Lagrange multipliers associated with the previous constraints, the first-order conditions for a $(t; C)$ solution of the maximization problem are

$$\forall i \qquad 0 = \alpha_i \frac{\partial V^i}{\partial \pi_F} - \mu \frac{\partial E^i}{\partial \pi_F} - v^i\left(-E - t^i \frac{\partial E^i}{\partial \pi_F}\right)$$

$$0 = \alpha_i \frac{\partial V^i}{\partial R} - \mu \frac{\partial E^i}{\partial R} - v^i\left(1 - t^i \frac{\partial E^i}{\partial R}\right). \qquad \text{(A8.2)}$$

the envelope theorem gives us that $(\partial V_i/\partial \pi_F)/(\partial V_i/\partial R) = -E_i$, and we obtain $t^i = \mu/\nu_i$ with $\nu_i = \alpha_i(\partial V_i/\partial R) = \alpha_i p_Q U^i_Q$.

This allows us to conclude for differentiated taxes, provided that the marginal utility of revenue is not the same for countries with different conditions (σ,ϕ), that, for example, $\partial^2 V/\partial\sigma\partial R = -(\pi_E/\pi_Q)\,\Phi_{\sigma L} l_R < 0$.

References

1. Barker, T. "Taxing Pollution instead of Employment: Greenhouse Gas Abatement through Fiscal Policy in the U.K." Discussion Paper No. 9, Economic and Social Research Council, Department of Applied Economics, Cambridge University, UK Energy-Environment-Economy Modelling.
2. Bohm, P. "Should Marginal Carbon Abatement Costs Be Equalized across Countries?" Research Papers in Economics, University of Stockholm, September 1993.
3. Chichilnisky, G. "The Abatement of Carbon Emissions in Industrial and Developing Countries." Invited presentation at the International Conference on the Economics of Climate Change, OECD/IEA, Paris, June 14–16, 1993. Published in *The Economics of Climate Change,* ed. Tom Jones (Paris: OECD, 1994), pp. 159–70.
4. Chichilnisky G., and G. Heal. "Who Should Abate Carbon Emissions? An International Viewpoint." *Economic Letters* 44 (1994): 443–49.
5. Coppel J. "Implementing a Global Abatement Policy: The Role of Transfers." Paper presented at the International Conference on the Economics of Climate Change, OECD, Paris, June 14–16, 1993.
6. Dean, A. "The Costs of Cutting Carbon Emissions: Results from Global Models." Paper presented at the International Conference on the Economics of Climate Change, OECD, Paris, June 14–16, 1993.
7. Dessus, B., and F. Pharabod. "Jérémie et Noé, deux scénarios énergétiques mondiaux à long terme." *Revue de l'énergie,* no. 421, June 1990.

8. Gille, B. Histoire des techniques. Encyclopédie de la Pléiade. Paris, NRF, p. 1652.
9. Godard, O. "Taxes." In *International Economic Instruments and Climate Change.* Paris: OECD, 1993, pp. 45–101.
10. Godard, O., and O. Beaumais. "Economie, croissance et environnement. De nouvelles stratégies pour de nouvelles relations." *Revue économique* 44, H.S. "Perspectives et réflexions stratégiques à moyen terme," pp. 143–76.
11. Goulder, L. H., and S. H. Schneider. "The Costs of Averting Climate Change: A Technological Bias in Standard Assessments." Working paper, Stanford University, July 29, 1995.
12. Hoeller, P., and J. Coppel. "Energy Taxation and Price Distorsions in Fossil Fuel Markets: Some Implications for Climate Change Policy." In *Climate Change, Designing a Practical Tax System.* Paris: OECD, 1992.
13. Hourcade, J. C. "Modelling Long-Run Scenarios: Methodology Lessons from a Prospective Study on a Low CO_2 Intensive Country." *Energy Policy* (March): 309–26.
14. Hourcade, J. C., and T. Chapuis. "No-Regret Potentials and Technical Innovation: A Viability Approach to Integrated Assessement of Climate Policies." *Energy Policy* 23, no. 4/5: 433–45.
15. Hourcade, J. C., R. Richels, and J. Robinson. "Estimating the Costs of Mitigating Greenhous Gases." In IPCC report, working group 3, 1995.
16. Jaffe, A. B., and R. N. Stavins. "The Energy Efficiency Gap. What Does It Mean?" *Energy Policy* 22, no. 10 (1994): 804–10.
17. Krugman, P. "Europe Jobless, America Penniless?" *Foreign Policy* (summer 1994): 19–34.
18. Laffont, J. J. *Effets externes et théorie économique.* Paris: Editions du CNRS, 1977.
19. Manne, A. S., and R. G. Richels. *Buying Greenhouse Insurance: The Economic Costs of* CO_2. Cambridge: MIT Press, 1992.
20. OECD. *"Global Warming: Economic Dimensions and Policy Responses."* Paris: OECD, 1995.
21. Sandmo, A. "Optimal Taxation in the Presence of Externalities." *Swedish Journal of Economics* 77 (1975): 86–98.
22. Walz, R., and M. Schon, et al. Gesamtwirtschaftliche Auswirkugen von Emissiosminderungsstrategien, Study for the Enquete Commission Protecting the Earth Atmosphere. Karlsruhe/Berlin: German Bundestag, 1994.

Chapter 9

Efficiency and Distribution in Computable Models of Carbon Emission Abatement

Joaquim Oliveira Martins
Peter Sturm

9.1 Introduction

Although much uncertainty surrounds the precise links between carbon emissions and their effect on climate, the risks involved are by now considered sufficiently large for the global community to have started discussing active policy measures. In this context special attention is being paid to the reduction of carbon emissions from the use of fossil fuels. The need for abatement action being generally recognized, the search is on for "efficient" policy instruments, that is, instruments that achieve a given abatement objective at minimum cost. In this context uniform global emission taxes and tradable emission quotas have been suggested as policy instruments of choice.

The initial consensus relating to the efficiency characteristics of a uniform global carbon tax and/or a system of tradable emission quotas has been challenged by Chichilnisky [5] and Chichilnisky and Heal [7], in which the authors (hereafter CH) claim that given the public goods character of emission abatement, a uniform emission tax or tradable emission quotas do not necessarily (in fact not usually) lead to Pareto-efficient outcomes, and that in the context of emission abatement policy efficiency and income distribution issues are intertwined; that is, the fundamental proposition of welfare economics that equity and efficiency are "orthogonal" (i.e., independent of each other) does not

The opinions expressed here are those of the authors and cannot be held to represent the views of the OECD or the IMF.

apply. The nonseparability of equity and efficiency issues is claimed to have important implications for the design of global carbon abatement policies and the choice of instruments to enforce it.

The separability of efficiency and equity is an underlying assumption in most of the computable general equilibrium (CGE) models that have hitherto been used to assess the economic costs of international agreements to reduce carbon emissions. For this reason the results obtained by CH have generated a debate on both the analytical correctness of the argument and its precise policy implications. This chapter aims at clarifying the analytical issues that determine cost efficiency in usual CGE abatement models. In this context some simulation results obtained with the OECD GREEN model are provided. Then it is shown under what conditions the equalization of marginal abatement costs across regions is not a necessary condition for achieving a Pareto-efficient allocation of scarce world resources and in what sense equity and efficiency issues cannot be separated. The consequences of these results for policy are briefly discussed.

9.2 Abatement Cost Models

This section recalls the efficiency conditions in the CGE models that do not embody environmental assets in the utility function (e.g., the OECD's GREEN model).[1] These models were designed with one specific aim: to assess the economic costs of reducing carbon emissions by a given amount determined exogenously. They are not concerned with the joint optimization of output and carbon emission abatement. In particular, they were not intended to evaluate the benefits from a reduction of carbon emissions.

9.2.1 Marginal Abatement Costs — To replicate in a simple way the typical structure of a CGE model, we assume two goods in a given economy: a carbon-free good C and a carbon-based good F, say, fossil fuels, which generates emissions of carbon dioxide E when consumed. The optimization problem of maximizing welfare under a given emission constraint can be formulated as follows:

$$\left.\begin{aligned} &\max\ U(C, F) \\ &\text{subject to}\quad g(C, F) = 0, \\ &h(F) = E \leq \overline{E} \end{aligned}\right\}, \tag{9.1}$$

[1] See Burniaux, Nicoletti, and Oliveira Martins [3].

where $g(\cdot)$ represents the production frontier and $h(\cdot)$ is the emission generation function associated with fossil-fuel consumption (with $h'(\cdot) > 0$) and $\overline{E}$ is the emission constraint. Under the normal convexity-concavity assumptions, the first-order conditions characterize the optimum:

$$\frac{\partial U}{\partial C} = \theta \cdot \frac{\partial g}{\partial C} \tag{9.2}$$

and

$$\frac{\partial U}{\partial F} = \theta \frac{\partial g}{\partial F} + [\lambda \cdot h'] = p_F + t_F, \tag{9.3}$$

where θ and λ are, respectively, the Lagrange multipliers associated with the resource and the emission constraint. Relation (9.3) says that the marginal social valuation (or the "correct" price) of F is equal to the competitive market price[2] (p_F) plus a term reflecting the valuation of the emission externality. In this expression the second right-hand term (in brackets) can be interpreted as the excise tax on fossil-fuels (t_F) needed to bring the private cost of F to its social cost. The excise tax t_E to be levied on carbon emissions is then equal to[3]

$$t_E = t_F \cdot \frac{1}{h'} = \lambda. \tag{9.4}$$

Therefore, the carbon tax is equal to the multiplier associated with the carbon constraint and can be interpreted as the marginal social (dis)utility of emissions. Under certain conditions at the social optimum this will equal the marginal abatement cost (MAC).[4] Bohm [2] suggested that marginal abatement costs should be defined in this way; CH have used another definition that might have created some confusion in the interpretation of their results.[5] They define the MAC as the opportunity cost of a unit of abatement in terms of consump-

---

[2]The competitive price of each good is equal to the shadow price of the resource constraint times the opportunity cost of production (see Varian [22]).

[3]Note that, by definition, $t_F \cdot dF = t_E \cdot dE$.

[4]By the envelope theorem, $dU/d\overline{E} = \lambda$.

[5]However, for reasons that will become clear shortly, the framework set up by Bohm [2] did not really clarify the debate because it is equivalent to the CH model with unlimited transfers among regions, implying the equalization of MACs under both CH and the standard definitions (see Chichilnisky and Heal [6]).

tion forgone. In our framework the CH definition of MAC would correspond to the trade-off between the consumption of the carbon-based good consumption and carbon abatement:[6]

$$\frac{dF}{d(-E)} = -\frac{1}{h'}. \tag{9.5}$$

Formulation (9.5) does not correspond to the standard definition of marginal abatement costs embodied in CGE models, even if, for presentational purposes, average abatement costs are often expressed in terms of gross domestic product (GDP) or consumption losses for a given level of abatement. Moreover, the emission generation functions h are typically different for each country (e.g., each fossil fuel mix has a different carbon content per unit of energy). The functions h would therefore have to be adjusted before there was any presumption that these opportunity costs should be equalized for Pareto efficiency.

9.2.2 Abatement Efficiency and Pareto Efficiency — Assume that there is a group of countries $i = 1, ..., n$, each applying an emission constraint such that

$$\sum_i \overline{E}_i = \overline{E}_W. \tag{9.6}$$

Therefore, the global emission target is reached by an emission constraint in each country. For example, the stabilization of carbon emissions in the OECD group is attained by stabilizing emissions in each country individually. Within this framework the question of cost-effectiveness can be raised; that is, is there a way of achieving the same global abatement at a lower cost? To simplify assume that two countries j and k are similar in every respect except for the emission generation function. Also suppose that country k generates (at the margin) more emissions per unit of energy than country i, that is,

$$h'_i < h'_k. \tag{9.7}$$

It is obvious that for the same excise tax on fossil fuels the induced marginal reduction in emissions is higher in country k than in country i. Therefore, instead of reducing consumption of fossil fuels at home, country i (the "high-cost/low-carbon" country) will be better-off to "buy" the corresponding amount of emission abatement in country k (the "low-cost/high-carbon" coun-

[6]Note that the CH model has only one consumption good.

try) and compensate this country for the costs incurred up to the point at which marginal abatement costs are equalized in the two countries. This efficiency gain could be extended to n countries, and from that it can be derived that in a cost-effective scheme, marginal abatement costs should be equalized.[7] The most simple way to implement this principle is to impose a global carbon emission constraint. It is precisely in this way that "cost-efficient" agreements are implemented in CGE models:

$$\begin{aligned} &\max\ U_i(C_i,\ F_i) \qquad\qquad \text{for } i = 1,\ \ldots,\ n, \\ &\text{subject to } g_i(C_i,\ F_i) = 0 \text{ and } \Sigma\, h_j(F_j) \leq \overline{E}_W. \end{aligned} \tag{9.8}$$

It follows immediately from the first-order conditions of this problem that

$$\frac{t_{F_1}}{h_1'} = \frac{t_{F_2}}{h_2'} = \ldots \frac{t_{F_n}}{h_n'} = \lambda, \tag{9.9}$$

where λ is the Lagrange multiplier associated with the (common) carbon constraint. As previously, this multiplier can be interpreted as the marginal abatement costs or the uniform tax levied on emissions in all countries. Whereas the tax on carbon emissions is equalized across countries, the excise taxes on consumption of F are country specific because they are tied to the characteristics of the emission generation functions, which can and do vary across countries.

This overall efficiency improvement might entail an extremely uneven distribution of the burden sharing across countries. This point is illustrated in table 9.1, which provides the simulation results with the OECD GREEN model of an international agreement to reduce world emissions by an amount corresponding to the stabilization of carbon emissions in the so-called Annex 1 group (i.e., OECD, eastern Europe, and the former Soviet Union). The comparison between the first and the second column in the table shows the efficiency gains from imposing an OECD-wide uniform carbon tax instead of a country- or region-specific tax. The average income losses in the OECD are reduced by roughly 0.1 percentage points over the period 1990–2050. At the world level the change in income losses is in the same order of magnitude. However, if the agreement is enlarged to the group of the so-called major emitters (i.e., Annex 1 plus China and India), the same global level of abatement can be achieved with a much lower world income loss (0.22% instead of

[7]Note again that this does not imply that the marginal productivity of abatement, h', needs to be equalized across countries.

Table 9.1

Distribution of gains and losses under different agreements.
Abatement scenario: reduction of world emissions corresponding to the stabilization of missions in Annex 1 countries at their 1990 levels.

Regions	**Unilateral taxes in OECD**	**Uniform tax: OECD**	**Uniform tax: Annex 1 + China + India**
OECD	−0.85	−0.76	−0.25
Annex 1	−0.86	−0.77	−0.20
China	−0.52	−0.47	−1.19
India	−0.07	−0.07	−0.74
Energy exporters	−3.62	−3.32	0.07
World	−1.07	−0.97	−0.22

Note: Average annual real income losses for the period 1990–2050 (as a percentage of deviation relative to the baseline scenario).
Source: GREEN model (OECD [17]).

0.97%). From table 9.1 it can be seen that this overall improvement leads to a disproportionate increase of the burden borne by "low-cost" countries (i.e., China and India). Interestingly, the major gainers from this abatement efficiency improvement are the energy-exporting countries.[8]

To secure the transfer of the emission abatement effort from high- to low-abatement-cost countries, it might be necessary to make transfers that compensate the latter for their incremental costs. Nonetheless, provided that the emission constraint is applied at the global level, it can be shown that abatement efficiency ensures Pareto efficiency and reciprocally.[9] In this context the issues of efficiency and equity are perfectly separable. This point is especially important for the design of a system of tradable permits,[10] as it implies that any initial distribution of allocation of permits will achieve efficiency. The considerations

[8]The intuition behind this result is the following: Because the overall resource allocation is optimized, there is a lower decrease of world energy consumption per unit of abatement. In addition, at the world level there is a shift from high-carbon domestic energy sources (typically coal) toward lower carbon imported ones (oil and gas). The lower reduction in energy demand and the substitution effect tend to increase the revenues of the energy-exporting countries.

[9]Indeed, a Pareto-efficient outcome will be characterized by the following program:

$$\max\ U_i(C_i, F_i)$$

$$\text{subject to}\quad U_k(C_k, F_k) \geq \overline{U}_k \quad \text{for } k, i = 1, \ldots, n \quad \text{and} \quad k \neq 1 \quad \text{and} \quad f_k(C_k, F_k) = 0$$

$$\text{and} \quad \sum_j h_j(F_j) \leq \overline{E_W}.$$

This would yield similar results to (9.8).

[10]Abstracting from uncertainty or transaction costs considerations.

Table 9.2

Distribution of gains and losses under different permit allocation rules. Abatement scenario: reduction of world emissions corresponding to the stabilization of emissions in Annex 1 countries at their 1990 levels.

Allocation rules	OECD	Annex 1	China and India	World
Initial quotas				
Grandfathering	56.5	84.7	15.3	100.0
Egalitarian	26.0	38.9	61.1	100.0
Losses/gains				
Grandfathering	−0.3	−0.1	−1.7	−0.2
Egalitarian	−0.7	−0.7	2.0	−0.2

Note: Initial quotas expressed as percentages of world emissions. For losses/gains, data are average income losses for the period 1990–2050 (as a percentage deviation relative to the baseline scenario).
Source: GREEN model (OECD [16]).

about income distribution can be viewed as a separate problem that can be solved, say, through a negotiation process. A quantified example of this remarkable property is shown in table 9.2. In the simulations presented, the same global abatement target as in the previous experiment is achieved by a system of permit trading with two extreme endowment rules: (1) a grandfathering rule, whereby countries/regions are endowed with emission quotas corresponding to their emissions in 1990, and (2) an egalitarian rule, whereby quotas are allocated in proportion to country/region population shares in 1990. Obviously, the second rule is more favorable to countries such as China and India and results in significant income gains in these countries compared with the losses incurred under the first rule.

Notwithstanding, the average world income losses remain exactly the same whatever the endowment rule is. It might happen in some cases that small differences appear between scenarios having the same abatement target but different permit allocations. This can be caused either by the approximate numerical solution provided by the resolution algorithm or by the different dynamic adjustment paths between scenarios. The nonseparability between equity and efficiency it is not what causes the gap.

9.3 Optimal Abatement Models

Ideally, instead of imposing an emission constraint, the level of global carbon emissions should be set at the (global) welfare-optimizing level. This means that each country or the world community as whole should be able to determine the effective damages of climate change and in this way establish a balance between costs and benefits of a policy action aiming to reduce the risk of global warming. Given the uncertainty surrounding the causal link between emissions, climate change, and its impacts on the economic system, this assessment requires an amount of information that is not currently available. Nonetheless, this is the research agenda of the so-called integrated assessment projects.[11]

The implications of considering the abatement externality directly in the utility function are profound because carbon emissions can be viewed as a public "bad" that is produced in a decentralized way by private consumption activities. This point was highlighted in Chichilnisky [5] and Chichilnisky and Heal [7].[12] Defining an objective function having global emissions as an argument implies that each country's utility function depends on the level of consumption of the carbon-based good in all the other countries:

$$U_i(C_i, F_i; E_w) \quad \text{with } E_w = \sum_j h_j(F_j). \tag{9.10}$$

9.3.1 The General Case: Country-Specific Production Frontiers — A Pareto optimum can be obtained by maximizing the utility of each country subject to the constraints on the utility levels of other countries and their specific production frontiers (as discussed shortly, the latter assumption is especially important):

$$\max\ U_i(C_i, F_i; E_w(F_1, F_2, \ldots, F_n)) \quad \text{subject to} \tag{9.11}$$

$$U_k(C_k, F_k; E_w) \geq \overline{U}_k \quad \text{for } k \neq i \quad \text{and} \quad g_k(C_k, F_k) = 0 \quad \text{for } k = 1, \ldots, n.$$

Using (9.10) and differentiating the corresponding Lagrangian with respect to all C_i and F_i, the first-order conditions for a given country i are

$$\mu_k \frac{\partial U_k}{\partial C_k} = \theta_k \cdot \frac{\partial g_k}{\partial C_k} \tag{9.12}$$

and

$$\mu_k \cdot \frac{\partial U_k}{\partial F_k} = (\theta_k \cdot \frac{\partial g_k}{\partial F_k}) + \left(\sum_j \mu_j \cdot \frac{\partial U_j}{\partial E_w}\right) \cdot h_k' \tag{9.13}$$

for $k = 1, \ldots, n$ and $\mu_i = 1$.

[11] The first applied models of this kind were built by Nordhaus [16] and Peck and Teisberg [18]. Several integrated assessment projects are currently under way (see, e.g., the second-generation model of Edmonds et al. [10] and, more recently, Prinn et al. [20] and Chichilnisky et al. [9]).

[12] See also Chichilnisky, Heal, and Starrett [8], and Hourcade and Gilotte [13].

Equation (9.13) can be interpreted much in the same way as the relation (9.3); that is, in each country the excise tax on fossil fuel consumption will be country specific and equal to $(T \cdot h'_k)$. However, the tax on carbon emissions T will be the same in each country and equal to

$$T = \sum_j \left(\mu_j \cdot \frac{\partial U_j}{\partial A} \right). \tag{9.14}$$

Moreover, from condition (9.13) it can also be shown that Pareto efficiency can be obtained only if

$$\frac{\mu_k \cdot \dfrac{\partial U_k}{\partial F_k} - \left(\theta_k \dfrac{\partial g_k}{\partial F_k} \right)}{h'_k} = T \qquad \text{for } k = 1, \ldots, n. \tag{9.15}$$

The conditions (9.14) and (9.15) entail two important departures from the previous results. First, in this framework the equality between the optimal carbon tax and the marginal abatement costs (or the marginal disutility of emissions) by country does not hold anymore. Indeed, from (9.14) the carbon tax corresponds now to a weighted sum of marginal abatement costs across regions.[13] In other words, although all countries face the same carbon tax, marginal abatement costs are not necessarily equalized across countries. This point was a source of confusion when interpreting the CH results because in their original paper the expression for the carbon tax was never made explicit (they refer only to the nonequalization of the MACs). Conversely, the equalization of marginal abatement costs across countries would require a system of differentiated carbon taxes, and this would correspond to the so-called Lindhal solution (see Foley [12]). It should be stressed that even if the equalization of marginal abatement utilities is not an objective per se, a *uniform* carbon tax is required for achieving cost efficiency. In this respect, our conclusion is different from Chichilnisky and Heal (1994).

Second, one must choose the appropriate set of multipliers or welfare weights in order to verify condition (9.15). Given that countries may differ in their preferences toward abatement and in their production conditions, not all combinations of utility weights will lead to Pareto efficiency. Typically, there

---

[13]This corresponds to the usual solution of the optimal tax with externalities (see Baumol and Oates [1]).

would be a Pareto point instead of a Pareto frontier, as is the case with only private goods.[14] This implies that the issues of equity and efficiency cannot be separated anymore (see also Laffont [14]).

Consequently, the delicate problem of the appropriate welfare weights becomes crucial for efficiency. For each simultaneous choice of welfare weights and the corresponding carbon tax—which could, for example, be the outcome of an international negotiation process—there will be an optimal level of global carbon abatement.[15]

It should be noted that the optimal solution depends on the actual preferences, income levels, and production characteristics. Gathering such an information set is a daunting task, but in the context of a CGE model, where a utility function similar to (9.10) is used, all the necessary information will be available by assumption. Such a model could then be used to run simulations illustrating how sensitive the results are to different choices of preferences toward the public good, forms of the production functions, and so on. For example, it would be interesting to analyze how the global abatement level depends on the different sets of welfare weights.

There is also a question of what interpretation should be given to the welfare weights. Formally, they correspond to the marginal valuations of the utilities of the different countries in a world welfare function. For a globally negotiated emission level, the weights will reflect the bargaining power of each region in the negotiations. It might also be interesting to relate the weights to the initial allocation of permits in a system of tradable permits. This would have strong implications for the design of a tradable permit scheme, as it would imply that only one initial permit allocation would be Pareto efficient for each level of global emissions.

9.3.2 Special Case: A Global Production Frontier — Chichilnisky and Heal [7] showed that a situation in which marginal abatement costs (in the sense of marginal consumption forgone by unit of abatement) will be equated across countries is one in which lump-sum transfers among countries can be realized without any limitation.[16] In our framework the possibility for unlimited trans-

---

[14]The set of relations (9.14)–(9.15) provide a system of linear $(n+1)$ equations determining *jointly* the optimal carbon tax T and the set of n multipliers μ_k.

[15]See Eyckmans, Proost, and Schokkaert [11], who, in the context of a numerical simulation model, showed that the optimal level of world abatement increases with the degree of aversion for income inequality.

[16]Noteworthy, allowing for international trade and especially trade in emission permits would not solve the problem of nonseparability. Indeed, international trade can replicate a situation of an integrated world economy only under first-best conditions.

fers would be equivalent to imposing a unique (global) production frontier in the equation (9.11), as follows:

$$\left.\begin{array}{rl} & \max\ U_i(C_i,\ F_i;\ E_w(F_1,\ F_2,\ \ldots,\ F_n)) \\ \text{subject to} & U_k(C_k,\ F_k;\ E_w) \geq U_k \quad \text{for } k \neq i \\ \text{and} & g_k\left(\sum C_k,\ \sum F_k\right) = 0 \end{array}\right\}, \tag{9.16}$$

and the first-order conditions for this problem are now

$$\mu_k \frac{\partial U_k}{\partial C_k} = \theta \cdot \frac{\partial g}{\partial C} \tag{9.17}$$

and

$$\mu_k \frac{\partial U_k}{\partial F_k} = \theta \cdot \frac{\partial g}{\partial F} + \left(\sum_j \mu_j \cdot \frac{\partial U_j}{\partial E_w}\right) \cdot h_k'$$

$$\text{for } k = 1,\ \ldots,\ n \quad \text{and} \quad \mu_i = 1, \tag{9.18}$$

where C and F correspond to the total (world) consumption level of the two goods. By replacing the welfare weights derived from (9.17) into (9.18) and simplifying, one gets

$$\frac{1}{n}\sum_k \left[\frac{\frac{\partial U_k}{\partial F_k}}{\frac{\partial U_k}{\partial C_k}}\right] - \frac{1}{n}\sum_j \left[\frac{\frac{\partial U_j}{\partial E_w}}{\frac{\partial U_j}{\partial C_j}}\right] \cdot \sum_k h_k' = \frac{\frac{\partial g}{\partial F}}{\frac{\partial g}{\partial C}}, \tag{9.19}$$

$A_k = -h_k(F_K)$. In this case the multipliers disappear from the optimality conditions. Expression (9.19) corresponds to the usual Lindhal-Bowen-Samuelson condition for the optimum with public goods. The sum of the marginal rates substitution are equal to the marginal rate of transformation between consumption and the carbon-based good.

9.4 Further Research and Conclusions

Contributing to this debate, Manne [15] referred to a case in which the externality (carbon emissions) originates in the production rather than in the utility function. In that case equity and efficiency are separable. Sturm [21] argues that in the context of international negotiations on climate change policies, only the (limited) notion of "efficiency in production" is operationally relevant. He showed that for this concept the distinction between public and private goods is irrelevant as long as there is a well-defined opportunity cost of regional abatement in terms of the private good, equivalent to the definition of marginal rate of transformation between private goods.

Prat [19] suggests that a constant-ratio mechanism (a ratio meaning a proportional division of the total emission quotas between countries) could separate the issues of equity and efficiency; once the ratio is determined, the (unique) optimal level of abatement can be decided by a planner. However, the implementation of decentralized procedure could raise serious practical problems. Another approach was put forward by Chao and Peck [4], who proposed a set of numerical simulations by which it is shown that the world optimal level of carbon abatement is not very sensitive to the income transfers among the countries. It goes without saying that the latter result depends crucially on the parameter calibration of the model.

These approaches adopt a somewhat pragmatic view of the problem that could be justified given the lack of information concerning the impacts of the climate change. Indeed, at this stage the joint optimization of income and emissions seems an exceedingly ambitious objective. Ultimately, the questions of equity have to be dealt with in the context of international negotiations by taking into account both net transfers or emission quota allocations.

References

1. Baumol, W. J., and W. E. Oates. *The Theory of Environmental Policy.* Cambridge: Cambridge University Press, 1988.
2. Bohm, P. "Should Marginal Carbon Abatement Costs be Equalised across Countries?" University of Stockholm Research Papers in Economics No. 1993: 12 WE, 1993.
3. Burniaux, J.-M., G. Nicoletti, and J. Oliveira Martins. "GREEN: A Global Model for Quantifying the Costs of Policies to Curb CO_2 Emissions." OECD *Economic Studies,* no. 19 (winter 1992).
4. Chao. H., and S. Peck. "Optimal Environmental Control and Distribution of Cost Burden for Global Climate Change." Mimeograph, Electric Power Research Institute, Palo Alto, California, 1995.
5. Chichilnisky, G. "Commentary on Implementing a Global Abatement Policy: The Role of Transfers." Paper presented at the Conference on the Economics of Climate Change, OECD/IEA, Paris, 1994.
6. Chichilnisky, G., and G. Heal. "Efficient Abatement and Marginal Abatement Costs." Mimeograph, 1993.
7. Chichilnisky, G., and G. Heal. "Who Should Abate Carbon Emissions?

An International View Point." *Economic Letters* 44 (spring 1994): 443–49.

8. Chichilnisky, G., G. Heal, and D. Starrett. "Equity and Efficiency in International Emission Permit Markets." Discussion paper, Stanford University, 1993. (Chapter 3 of this volume)
9. Chichilnisky, G., V. Gornitz, G. Heal, D. Hind, and C. Rosenzweig. "Building Linkages among Climate Impacts and Economics: A New Approach to Integrated Assessment." Working paper, Global Systems Initiative, Columbia University, June 1996.
10. Edmonds, J. A., H. M. Pitcher, N. J. Rosemberg, and T. M. L. Wigley. "Design for the Global Climate Assessment Model." Mimeograph, 1993.
11. Eyckmans, J., S. Proost, and E. Schokkaert. "Efficiency and Distribution in Greenhouse Negotiations." *Kyklos* 46 (1993).
12. Foley, D. "Lindhal's Solution and the Core of an Economy with Public Goods." *Econometrica* 38, no. 1 (1970): 66–72.
13. Hourcade, J.-C., and L. Gilotte. "Some Paradoxical Issues about an International Carbon Tax." Paper presented at the Annual Conference of the European Association of Environment and Resource Economics, Dublin, 1994.
14. Laffont, J. J. (1989), Fundamentals of Public Economics, The MIT Press, Cambridge, MA.
15. Manne, A. "Greenhouse Gas Abatement—Toward Pareto-Optimality in Integrated Assessments." Mimeograph, School of Engineering, Stanford University, 1993.
16. Nordhaus, W. D. "The DICE Model: Background and Structure of a Dynamic Integrated Climate-Economy Model of the Economics of Global Warming." Cowles Foundation Discussion Paper No. 1009, 1992.
17. OECD. *Global Warming: Economic Dimensions and Policy Responses.* Paris: OECD, 1995.
18. Peck, S. C., and T. J. Teisberg. "CETA: A Model for Carbon Emission Assessment." *The Energy Journal* 13, no. 1.
19. Prat, A. "Efficiency Properties of a Constant-Ratio Mechanism for the Distribution of Tradable Emission Permits." Mimeograph, Stanford University, 1995. (Chapter 6 of this volume).
20. Prinn, R., et al. "Integrated Global System Model for Climate Policy Analysis." MIT Joint Programme on the Science and Policy of Global Change, Report No. 7, June 1996.
21. Sturm, P. "The Efficiency of Greenhouse Gas Emission Abatement and International Equity." Massey University Discussion Paper No. 95.9, June 1995.
22. Varian, H. R. *Microeconomic Analysis,* W. W. Norton, 1984.

Chapter 10
Securitizing the Biosphere

Graciela Chichilnisky
Geoffrey Heal

A handful of firms in traditionally dirty industries have decided that they can make more money by embracing environmental goals than by fighting them. At the leading edge of the environmental movement, British Petroleum, Monsanto, Dupont, Compaq, 3M, S.C. Johnson, Dow Chemical, Weyerhauser, and Interface are major corporations improving their financial performance by cleaning and greening their operations [1]. They are making money by reducing their environmental impact. This is not entirely surprising: Costanza et al. [2] have suggested that environmental services have great value, although they did not indicate how this value can be realized. Here we take this line of argument further and propose various economic instruments that would allow investors to obtain economic returns from environmental assets such as forests and landscapes while ensuring their conservation. One of us [3] has proposed the creation of an international financial institution to promote this process, a suggestion officially adopted by the Group of 77 developing countries and China in the Kyoto meetings on December 1, 1997.

The environment's services are clearly valuable. The air we breathe, the water we drink, and the food we eat are all available only because of services provided by the environment. How can we transform these values into income while conserving the underlying natural capital? We have to "securitize" natural capital and environmental goods and services and enroll market forces in their conservation. This means assigning to corporations—possibly innovative public-private corporate partnerships—the obligation to manage and conserve natural capital in exchange for the right to the benefits from selling the services provided. E. O. Wilson [5] talks of "the need to draw more income from the wildlands without killing them, and so to give the invisible hand of free market economics a green thumb." Privatizing natural capital and ecosystem services is a key step. It enlists self-interest and the profit motive in the cause of the environment. Although these motives will never conserve everything that we value in the environment, they will conserve a lot, leaving regulation and appeals to higher motives to focus on really hard cases.

10.1 Investing in the Biosphere

In 1996 New York City decided to invest between $1 billion and $1.5 billion in natural capital in the expectation of producing cost savings of $6 billion to $8 billion over 10 years, giving an internal rate of return of between 90% and 170% and a payback period of between four and seven years. This return is an order of magnitude higher than is normally available, especially on relatively riskless investments. New York's water comes from a watershed in the Catskill Mountains. Until recently, purification processes carried out by root systems and microorganisms in the soil as the water percolates through, together with filtration and sedimentation occurring during this flow, were sufficient to cleanse the water to the standards required by the U.S. Environmental Protection Agency (EPA). Recently, sewage, fertilizer, and pesticides in the soil reduced the efficacy of this process to the point that New York's water no longer met EPA standards. The city was faced with a choice: restore the integrity of the Catskill ecosystems, or build a filtration plant at a capital cost of $6 billion to $8 billion, plus running costs of the order of $300 million annually. In other words, New York had to invest in natural capital or in physical capital.

Which was more attractive? Investment in natural capital in this case meant buying land in and around the watershed so that its use could be restricted and subsidizing the construction of better sewage treatment plants. The total cost of measures of this type needed to restore the watershed is expected to be in the range of $1 billion to $1.5 billion. Thus, investing $1 billion to $1.5 billion in natural capital could save an investment of $6 billion to $8 billion in physical capital. These calculations are conservative, as they consider only one watershed service, although watersheds, typically forests, often provide other important services.

The support of biodiversity is one, and carbon sequestration is another. The commercial value of biodiversity can be partly captured by bioprospecting deals such as that between Merck and Costa Rica's InBio (see the following discussion). Joint implementation offers the possibility of commercializing the carbon sequestration role. This allows carbon emitters in industrial countries to be credited with emission reductions that they support financially in developing countries: It allows them to buy abatement credits through bilateral trades. Several such deals have been brokered by the Global Environment Facility. The implementation of a global multilateral carbon emission market, as proposed by the United States in the context of the Kyoto negotiations, will provide a more robust way of selling sequestration services: It will allow credits for carbon sequestration, which can be cashed in the emissions market. In principle, then, a forest ecosystem can sell many different services. Recent provisions in Costa Rica recognize this, as they credit forested conservation areas with income for the services that they provide as watersheds and as carbon sinks to the extent of $50 per hectare for the former and $10 per hectare for the latter. This is sufficient to tip the balance in favor of conserving land of marginal agricultural value.

Agriculture provides another example of the returns from investing in biodiversity to preserve genetic variation. In the early 1970s a virus called the "grassy stunt" virus posed a major threat to Asia's rice crop. This threat was defeated by the transfer of an immunity-conveying gene from wild rice to commercial varieties. A similar event occurred in 1976: Another threatening disease was defeated by transferring to commercial varieties the immunity carried by certain strains of wild rice, preserved for just this reason by the International Rice Research Institute in the Philippines. The returns to this investment in conservation have been almost incalculable.

10.2 Securitizing the Biosphere

New York City recently floated an "environmental bond issue" and will use the proceeds to restore the functioning of the watershed ecosystems responsible for water purification. The cost of the bond issue will be met by the savings produced: the avoidance of a capital investment of $6 billion to $8 billion, plus the $300 million annual running costs of the plant. The cash that would otherwise have gone to these will pay the interest on the bonds.

These cost savings could have been "securitized." This means pledging a fraction of them to the providers of the capital as a return on their investment. New York City could have opened a "watershed savings account" into which it paid a fraction of the costs avoided by not having to build and run a filtration plant. This account would then pay investors for the use of their capital.

This same financial structure is already used in securitizing the savings from increased energy efficiency in buildings. Securitization of the savings involves issuing contracts—securities—entitling their owners to a specified fraction of the savings. Typically, these contracts are tradable, issued to the providers of capital, and can be sold by them, even before the savings are realized. This is a way of making investment in saving energy attractive: It does not imply any transfer of ownership of the underlying asset. The U.S. Department of Energy has a standard protocol for estimating the savings from enhanced building energy efficiency. Several financial agencies are willing to accept these estimates of energy savings as collateral for loans.

Securitization of ecosystem services is not new. It has a pedigree going back at least to 1624, when in a deal structured by the Grand Duke Ferdenando II of Tuscany, a member of the Medici family, the Monte di Paschi di Siena[1] was recapitalized by the issue of bonds secured on the income received by the grand duke from the use of his pasture lands in the Maremma.

10.3 Privatizing the Biosphere

One could take the introduction of market forces a step further. Imagine a corporation managing the restoration of New York's watershed, the "Catskill Watershed Corporation." This has the right to sell the services of the ecosystem, which is different from ownership of the asset itself. In the case of New York's watershed, the services are the provision of water meeting EPA standards. Ownership of this right would enable the corporation to raise capital from capital markets, to be used for meeting the costs of conserving New York's watershed. If we were conserving biodiversity rather than a watershed, the corporation would own and sell (or licence) the rights to intellectual property derived from the biodiversity. Such a framework would harness private capital and market forces in the service of environmental conservation.

In privatizing the provision of ecosystem services, we are creating private property rights where none existed previously. Common property is being assigned to individuals and corporations. If the common property is in a privately produced public good, then the issues considered in the other chapters of this book are relevant: The property rights have to be assigned in particular ways to ensure that an efficient outcome is attained. In the case of a watershed, the purity of the water is indeed a local privately produced public good, as it is determined by pollution and land use decisions of a large number of people in the area of the watershed. Water quality is a public good, as it is a function of pollution, which is in turn clearly a public good (bad). The implications of this for the management of a watershed are set out in the Appendix of this chapter.

10.4 Financing the Biosphere

What is the practical potential of securitization and privatization? How significant a contribution could it make to meeting the challenge of conserving the biosphere?

---

[1]Monte di Paschi di Siena, founded in 1472, is currently one of the largest banks in Italy.

Many important watersheds are threatened by development. These include not only that of New York but also the watersheds of Rio de Janeiro, the basin of the river Paraibo do Sul in the Mata Atlantica coastal forest in Brazil, and the watershed for parts of Buenos Aires. The Mata Atlantica is a region of great biotic uniqueness, and its conservation would convey benefits far in excess of the value of the water provided. Thus, arrangements of the type discussed could be applied to the watersheds of some of the largest cities in the Western Hemisphere and undoubtedly many more. Within the United States alone, over 140 cities are now considering watershed conservation as an alternative to water purification. Not only could this be cost effective, but it could also represent a major impetus to environmental conservation and a happy alignment of market forces with the environment.

The EPA recently estimated that over the next 20 years, ensuring safe and adequate drinking water in the United States will require infrastructure investment of $138.4 billion. The equivalent figure worldwide will be in the trillions of dollars. Taken in the context of the other pressing infrastructure needs of developing countries, such a number is almost certainly not attainable by the public sector. Watershed conservation could cut the investment needed substantially, and securitization or privatization could ensure that much of the balance remaining is provided by the private sector.

What is the potential for application of privatization or securitization to ecosystems other than watersheds? Daily [4] identifies the following social and economic functions of ecosystem services: purification of air and water, mitigation of floods and droughts, detoxification and decomposition of wastes, generation and preservation of soils, control of the vast majority of potential agricultural pests, pollination of crops and natural vegetation, dispersal of seeds, cycling of nutrients, maintenance of biodiversity, protection of coastal shores from erosion, protection from harmful ultraviolet rays, partial stabilization of the climate, and provision of aesthetic beauty and intellectual stimulation that lift the human spirit.

Which of these are amenable to the approach that we have indicated? One clear prerequisite is that the ecosystem to be conserved must provide goods or services to which a commercial value can be attached. Watersheds satisfy this criterion: Drinkable water is becoming increasingly scarce, and indeed the availability of such water is one of the main constraints on health improvements in many poorer countries.

Commercial value of an ecosystem service is necessary but not sufficient for privatization. Some of that value has to be appropriable by the producer. A critical issue in deciding whether ecosystem services can be privatized is the extent to which they are public goods. Pure public goods are challenging to privatize; they are goods that if provided for one are provided for all. It is hard, although often not impossible, to exclude from benefiting from their provision those who do not contribute to their costs, so that their providers cannot appropriate all their returns. Water quality is a public good in the sense that if it is improved for one user of a watershed, then it is improved for all. However, the consumption of water itself is excludable, so the watershed case involves bundling a public with a private good. Knowledge, an intermediate category and one of the services of biodiversity, has to be commercialized with care, as shown by the need to protect it with patents, copyrights, and other supports of intellectual property rights.

An ecosystem service that could be treated by securitization or privatization is the support of ecotourism, which requires a significant degree of ecological integrity. It is natural to expect that private investment will be forthcoming to finance the conservation of a region with significant ecotourism potential in return for the right to some of the revenues. The growth of private game reserves is an obvious manifestation of this. There is a close economic resemblance to watersheds, in that the preservation of the ecosystems supporting ecotourism is a public good and benefits all. However, the hotel rooms and guide services are private goods whose value is enhanced by the public good.

The International Rice Research Institute played a key role in preserving access to genetic material that might provide immunity to disastrous new diseases. They played this role by conserving a wide range of rice strains, a clear indication of the commercial value of biodiversity. Costa Rica and the pharmaceutical company Merck have made an innovative financial deal aimed at appropriating to Costa Rica some of the economic value of its biodiversity. The deal has three parts: an agreement by Costa Rica to conserve an area of forest, supported by a payment from Merck; an agreement giving Merck access to the results of bioprospecting in this forest; and an agreement that Merck will pay Costa Rica a royalty on products developed from this bioprospecting. The deal represents a first step in providing a conservation agency in a developing country with a financial stake in the intellectual property of its biodiversity.

Is there a possibility of securitizing biodiversity as a way of encouraging private capital to conserve genetic variation and capture some of its commercial value? Genetic information has been securitized. Incyte, a biotechnology company, has as its only product a database of information about genetic structures. This information has been heavily processed; biodiversity in its natural state represents unprocessed genetic information, which is less commercially usable. There might be a role for private capital in establishing a "preprocessing" center for genetic information from developing countries. Such a center could conduct some preliminary analysis and then sell the right to use these, with a royalty to the originating country.

10.5 Conclusion

For certain types of ecosystem service, privatization or securitization represents a very real possibility. It could play a central role in realizing the economic value of the underlying asset and thus provide powerful economic incentives to conserve it for future generations. Examples of ecosystems services that might be privatized include watershed and carbon sequestration services, preservation of wild animals as a basis for ecotourism, and pollination. Biodiversity as a source of genetic knowledge is also a possible candidate for this treatment, although it is a case that presents more problems.

Appendix

Here we give a formal analysis of the issues discussed previously. This is presented in the context of the securitization and privatization of a watershed. The quality of the water in the watershed is assumed to be a public good, in that it is determined by the levels of polluting activities in the watershed region, and these are traditional public bads. We can also note that the quality of the water in a watershed is the same for all users of the watershed and is thus a nonexcludable property of the watershed. The framework we consider is as follows:

1. The right to sell water is owned by a private company, the Water Company, which sells water to individuals at a market-clearing price.
2. Individuals own shares in the Water Company and receive its profits as dividends.
3. The Water Company is responsible for ensuring that the level of pollution in the water is below a standard that is set by the government. The government is not explicitly modeled. Its only role is to set this standard.
4. Any individual wanting to emit effluent must own effluent permits for the appropriate amount. These permits are tradable and are initially distributed to individuals. The only mechanism by which the Water Company can ensure compliance with the government's standards is by buying effluent permits from individuals until the number left in individual ownership just matches the effluent targets. The Water Company thus has to invest in attaining the specified level of water quality, and it then sells water to individuals and buys permits from them. The difference is the profits, which are distributed as dividends to the owners (individuals).

We work with a simple formal model. Let $u_i(c_i, w_i, q)$ be the utility of a person consuming an amount c_i of a private consumption good and an amount

w_i of water of quality q. The quality, which is a function of the level of pollution in the watershed, is the same for all and thus is a public good for the community at issue. The function u is assumed to be smooth and quasi-concave. There are I such individuals. The only goods in the economy are water and the consumption good. Production of the consumption good leads to effluent, which reduces water quality, so that we can write $q_i = f_i(y_i)$, where y_i is the amount of the private good produced by person i and q_i is the resulting pollution. For simplicity, we do not distinguish people from firms. The production function f_i is assumed to be smooth, increasing, and strictly concave. The total amount of pollution in the water is $q = \Sigma_i q_i$, and this level is the same for all people who use the watershed. The level of pollution is thus a public good that is privately produced. The water itself is of course a private good, as what one person drinks another cannot. There is a fixed amount of water per time period; only the quality of the water can vary. Thus, in each period individuals' consumption levels of water sum to $w : \Sigma_{i=1}^{I} w_i = w$. We model a stationary equilibrium that is the same each period.

Efficient Allocations

The central issues in determining an efficient allocation of resources are now as follows. What should each person produce, what should each consume, and how should the water be allocated between people? Once we know what will be produced and by whom, we know the quality of the water through the functions f_i. Standard arguments show that any Pareto-efficient allocation can be characterized as one that maximizes the utility of one person subject to the constraints that

1. total effluent equals the sum of individual effluents, $q = \Sigma_{i=1}^{I} q_i$ and $q_i = f_i(y_i)$;
2. total consumption equals total production, $\Sigma_{i=1}^{I} c_i = \Sigma_{i=1}^{i} y_i$;
3. supply of and demand for water balance, $w = \Sigma_{i=1}^{I} w_i$; and
4. each other individual attains a specified utility level, $u_j(c_j, w_j, q) = \overline{u}_j \, \forall j \neq i$.

The specified utility levels are parameters in this problem. All Pareto-efficient allocations can be characterized as solutions to such a problem as these levels vary. Formally, this problem is

$$\max u_i(c_i, w_i, q) \quad \text{subject to} \quad q_i = f_i(y_i),$$

$$q = \sum_{i=1}^{I} q_i, \; w = \sum_{i=1}^{I} w_i \qquad \text{(A10.1)}$$

$$\sum_{i=1}^{I} c_i = \sum_{i=1}^{I} y_i \quad \text{and} \quad u_j(c_j, w_j, q) = \overline{u}_j \qquad \forall j \neq i.$$

The first-order conditions for optimality are easily shown to be

$$\frac{\partial u_i}{\partial c_i} + p_c = 0, \ \lambda_j \frac{\partial u_j}{\partial c_j} + p_c = 0, \tag{A10.2}$$

$$\frac{\partial u_i}{\partial w_i} + p_w = 0, \ \lambda_j \frac{\partial u_j}{\partial w_j} + p_w = 0, \tag{A10.3}$$

and

$$\frac{\partial f_i}{\partial y_i} \sum_{j=1}^{I} \lambda_j \frac{\partial u_j}{\partial q} - p_c = 0. \tag{A10.4}$$

These imply that

$$\frac{\partial f_i}{\partial y_i} = \frac{\partial u_i / \partial c_i}{\Sigma_{j=1}^{I} \lambda_j \partial u_j / \partial c_j}, \tag{A10.5}$$

which is a version of the Lindahl-Bowen-Samuelson result that the marginal rate of transformation between a public and a private good should equal the sum of the marginal rates of substitution between them across individuals.

Privatization and Securitization

The analysis so far is institution free. Equations (A10.2) and (A10.5) characterize Pareto-efficient allocations of water and consumption. Next we check whether or when the framework set out in this chapter will attain these conditions. A typical individual faces the problem

$$\max\ u_i(c_i, w_i, \overline{q}) \quad \text{subject to } p_w w + p_c c = p_c y_i + s_i \Pi + p_e \{T_i - q_i\}. \tag{A10.6}$$

Here $\overline{q}$ is the total level of pollution or of effluents permitted by the tradable permit system. In the budget constraint the market value of the individual's production $p_c y_i$ is supplemented by

1. his or her share of the profits of the Water Company (the profits are Π, and the agent's share in these is s_i), and
2. the market value of the person's net trade $p_e\{T_i - q_i\}$ in effluent permits. The term T_i is i's target level of effluent and q_i the actual level, and the difference is available for sale (if positive) or has to be bought (if negative) at the market price of a permit, p_e.

Applying standard techniques to (A10.6), we see that the conditions characterizing the individual's choice are

$$\frac{\partial u_i}{\partial w_i} + \mu_i p_w = 0, \tag{A10.7}$$

$$\frac{\partial u_i}{\partial c_i} + \mu_i p_c = 0, \tag{A10.8}$$

and

$$\frac{\partial u_i}{\partial c_i} = \mu_i p_e \frac{\partial f_i}{\partial y_i}, \tag{A10.9}$$

so that

$$\frac{\partial f_i}{\partial y_i} = \frac{p_c}{p_e}. \tag{A10.10}$$

We naturally want to know whether the market will lead to Pareto efficiency, so the next step is to check whether the individual choices as modeled previously above satisfy conditions (A10.2) to (A10.5) for Pareto efficiency, that is, whether conditions (A10.7) to (A10.10) describing people's choices imply the conditions (A10.2) to (A10.5) for efficiency. One difference is immediate from comparing (A10.5) with (A10.10):

$$\frac{\partial f_i}{\partial y_i} = \frac{\lambda_j \partial u_j / \partial c_j}{\Sigma_{k=1}^{I} \lambda_k \partial u_k / \partial c_k} \quad \text{vs.} \quad \frac{\partial f_i}{\partial y_i} = \frac{p_c}{p_e}.$$

The right-hand side of (A10.10) is independent of the index i, and the right-hand side of (A10.5) is not and, in principle, could be different for every different person. A necessary condition for the market solution via individual choices to be efficient is thus that $\partial u_i/\partial c_i$ and $\lambda_i \partial u_j/\partial c_j$ be the same for all i and j. This is a familiar condition from the analysis of chapter 3: we are seeking a condition on the equality of the marginal valuations of private consumption across individuals, which is a condition on the distribution of income. The market will attain efficiency only at those distributions at which this condition is satisfied. By following the arguments in Chichilnisky, Heal, and Starrett in chapter 3 of this volume, we can establish the following result:

PROPOSITION 1 Given a regularity condition stated in chapter 3 of this volume, the system of privatization and securitization based on effluent permits described above will lead to Pareto-efficient outcomes only if the distribution of effluent permit lies in a submanifold of codimension one of the sets of possible permit allocations.

Securitization and privatization can work to attain a Pareto-efficient outcome, but there are specific prerequisites that are necessary for this. In particular, the distribution of the property rights created by the privatization must satisfy certain conditions; that is, they cannot be distributed arbitrarily if the outcome is to be efficient.

References

1. *Business Week,* November 10, 1997, pp. 98–99.
2. Costanza, R. et al. "The Value of the World's Ecosystem Services and Natural Capital." *Nature* 387, no. 6230 (May 15, 1997).
3. Chichilnisky, G. "Development and Global Finance: The Case for an International Bank for Environmental Settlements." UNDP-UNESCO Office of Development Studies, New York, September 1996.
4. Daily, G. ed. *Nature's Services: Societal Dependence on Natural Ecosystems.* Washington D.C.: Island Press, 1997.
5. Wilson, E. O. *The Diversity of Life.* New York: W. W. Norton, 1993.

Chapter 11

Equity and Efficiency in Emission Markets: The Case for an International Bank for Environmental Settlements

Graciela Chichilnisky

11.1 Introduction

Global institutions created after World War II—the World Bank, the International Monetary Fund (IMF), and the General Agreement on Tariffs and Trade (GATT)[1]—led the world into an unprecedented period of industrialization, material expansion, and global commerce. Called the Bretton Woods institutions, they emerged from the premise that trade and economic growth could help defuse international conflicts and accelerate the reconstruction after the

The proposal for the International Bank for Environmental Settlements (IBES) was officially presented at the May 1994 Workshop on Joint Implementation organized with the support of the Global Environmental Facility (GEF) and the Framework Convention on Climate Change (FCCC) at Columbia University, in various FCCC meetings and at an invited address to the annual meetings of the World Bank in December 1995, see also Chichilnisky [4]. In the preparation of this proposal, I have benefited from the discussions of several members of the Intergovernmental Negotiating Committee (INC) of the FCCC, who provided important insights: Minister Raúl Estrada-Oyuela, chair of the INC/FCCC; H. E. Ismail Razali, ambassador, permanent mission of the Malaysian to the United Nations; Mr. Xialong Wang, third secretary, Chinese permanent mission to the United Nations; Mr. James Baba, deputy permanent representative of Uganda to the United Nations; and Dr. John Ashe, counsellor, permanent mission of Antigua and Barbuda to the United Nations. In addition to emissions trading, the proposal included the creation of an IBES, which could help reconcile efficiency and equity in emissions markets; the two key features are deeply connected in these type of markets, as shown in this chapter. This chapter originally was presented at a workshop organized by New York Law School at the Villa La Pietra in Florence in the summer of 1996. I thank the participants of the workshop, especially Richard Stewart, Stephen Breyer, and Richard Ravel, for valuable comments and suggestions.

[1] Among others. GATT is now the World Trade Organization (WTO).

devastations of war.[2] Under the aegis of these institutions, economic growth led to record industrial expansion and resulted in an ever-increasing use of energy and natural resources. At the end of this 50-year period, we face global environmental challenges that originate from the success of industrialization itself: For the first time in history, economic activity has reached levels at which it can alter[3] the atmosphere of the planet and the complex web of species that constitute life on earth. Humans have the ability to destroy in a few years the massive infrastructure that supports the survival of the human species, the global habitat to which humans have adapted optimally throughout the ages. Industrial societies' intensive use of the earth's resources is reaching its logical limits and is now under close scrutiny.

An international body, the UN Framework Convention on Climate Change (FCCC), is responsible for negotiating a response to the problems created by the rapidly increasing emission of greenhouse gases into the planet's atmosphere. Two aspects that play an important role in the climate negotiations are *efficiency* and *equity* in the use of the world's resources. Efficiency is crucial in a period in which we seek to reduce the use of resources, as adopting efficient measures can by itself lessen resource use without negative consequences. However, fairness is also key, as many of the environmental issues considered in the global negotiations (the rights to use the planet's atmosphere and the world's biodiversity) involve the use of global *public goods* and require international negotiations in which fairness plays an important role. There is no agreed way to reach a fair allocation in the use of the world's resources, yet without one it is difficult to visualize solutions that are both politically feasible and stable in the long run.

This chapter looks at the issues of equity and efficiency in the allocation of environmental resources and examines the global institutions that may be needed to implement solutions that are both equitable and efficient. It starts from the premise that environmental markets will play an important role in the allocation of environmental resources. Such markets already exist in some countries, for example, those to trade the rights to emit sulfur dioxide (SO_2) in the Chicago Board of Trade. Others are being created, such as water markets and the carbon dioxide (CO_2) markets provided by the Kyoto Protocol for the trading of carbon emission rights among Annex B nations. Carbon emission markets are based on the limitations on emissions for Annex B countries

[2]See G. Chichilnisky, "The Greening of Bretton Woods," *The Financial Times*, January 10, 1996, Business and Environmental Section, and C. Bernandes, "Environmental Assets and Derivatives," *Derivatives Week* 5, no. 22 (June 3, 1996).

[3]In many cases irreversibly.

agreed in the Kyoto Protocol and require specific developments before they can be implemented. Prior results in Chichilnisky, Heal, and Starrett [13] and chapter 3 of this book (hereafter referred to as the CHS chapter) show that traditional market approaches might not ensure efficiency. The CHS chapter proposes a global mechanism or institution to overcome this shortcoming, one that has sensitivities where the Bretton Woods institutions fell short. Called an International Bank for Environmental Settlements (IBES), the mechanism proposed could be stand-alone or part of other global institutions. Its overall role would be to offer financial incentives for economic progress that is harmonious with environmental conservation.[4] How the IBES could work in practice is the concern here.

The proposal for an IBES arises from the need to overcome the shortcomings of traditional markets in the environmental area: These do not ensure efficiency. The shortcomings arise from a somewhat unexpected connection between efficiency and the initial allocation of property rights in environmental markets, a connection that is not present in standard markets (see Chichilnisky, Heal, and Starrett [13] and the CHS chapter). Building on this connection, the present chapter goes further to show in the Appendix that in certain cases market efficiency requires a preferential treatment for lower-income groups. Here a new concept is introduced: the "manifold of efficient distributions of property rights," which is the set of initial allocations of rights from which the competitive market can achieve efficient solutions.

A preferential treatment of lower-income nations would give these countries proportionately more rights of use of the atmosphere as a global public good. It would in effect require that the first countries to abate carbon emissions should be the industrial nations. These were the same conclusions obtained theoretically in Chichilnisky [19] and Chichilnisky and Heal [10] and later accepted by 166 nations in the 1997 Conference of the Parties 3 (COP3) and the resulting Kyoto Protocol, which was recently signed by the United States. More recently, the conclusions were endorsed in the COP4 in Buenos Aires in November 1998. In all cases the commitments to abate carbon emissions are from the industrial nations (Annex B countries): a 5.2% decrease by the period 2008–12, representing a 30% drop from current projections. The Appendix offers further theoretical and empirical support for the desirability of this outcome.

It might be useful to point out that the economic principles discussed here apply to other environmental assets, such as biodiversity, water, soil, and for-

[4] See Chichilnisky [14].

ests. However, the examples and data provided in this chapter concentrate on the use of the atmosphere in the emission of greenhouse gases, mostly derived from the burning of fossil fuels—coal and petroleum—to generate energy. The Appendix provides runs of the GREEN/PIR global model that is different from that in the CHS chapter because it does not incorporate environmental quality in the utility function. Yet even in the GREEN/PIR model it can be seen that the runs that assign more rights to emit to lower-income nations have somewhat lower costs of meeting the emission reduction targets. A possible explanation is provided in the historical data analyzed in the Appendix: On average a dollar invested in developing nations has a higher return that the same dollar invested in industrial nations. To the extent that abating emissions leads to lower investment, for efficiency abatement should initiate in the industrial nations, as doing this minimizes the negative effect of a drop in investment. Theoretical results supporting these conclusions appear also, within a different model, in Chichilnisky [19] and Chichilnisky and Heal [11].

The overall role of the IBES must be seen in the context of promoting a new form of economic development that contrasts with the resource-intensive policies followed by the Bretton Woods institutions. The imperative suggesting a real change in the use of resources appears clear enough. There is in addition a global economic trend that could ease the transition to a society that is more conservative in the use of resources: Industrial society is in the process of transforming itself into a knowledge society. This transformation, which has been called the "knowledge revolution" [5] is acquiring a global reach. The new economy that emerges is not a service economy as previously thought, as it involves mostly highly skilled labor. Through this ongoing transformation humans could achieve a new form of economic organization in which the most important input of production is no longer machines but human knowledge. Instead of burning fossil fuels to power machines, we could burn information to power knowledge. Information is a much cleaner fuel than coal or petroleum and can put humans rather than machines at the center of economic progress, leading to a knowledge-intensive rather than a resource-intensive form of growth.

11.1.1 IBES: A Two-Sided Coin to Overcome the North-South Divide — This chapter seeks to explain why a new mechanism or institution (IBES) might be required to complement the Bretton Woods institutions, how this

[5] A concept introduced and studied in Chichilnisky [5,6,8]; the expression itself is a trademark of the author.

would work in practice, and why its role would complement environmental markets but go further than anything that unaided markets can achieve. The global financial mechanism proposed here would exceed that of a standard market for trading emission rights. A standard market—for example, a stock market—trades private goods, and when private goods are traded, competitive markets are efficient independently of the allocation of property rights. However when *public goods* are traded, competitive markets might not reach efficient outcomes without an appropriate distribution of initial user rights, which are also called property rights (CHS chapter). The *manifold of user rights* from which efficient allocations could be reached is defined in the Appendix; it shows that in certain cases lower-income regions should be assigned a larger share of the use of the global commons in order to reach efficiency. Within the global climate negotiations, this means that developing countries should be assigned proportionately more user rights on global environmental assets, such as the planet's atmosphere, to ensure efficient market solutions. From these findings it follows that new global institutions, such as the proposed IBES, would be needed to complement the Bretton Woods institutions in order to implement global emissions markets and ensure their efficiency. For example, the IBES could help to negotiate global user rights or a basic borrowing and lending rate[6] or to help to establish property rights.[7]

The IBES would be like a two-sided coin, in that it would combine market-based instruments with political mechanisms. The latter would involve political representation and would give all nations effective participation, providing a continuous role of the type that the FCCC plays today in yearly meetings. This type of participation is more congenial to developing nations that are not comfortable with financial markets. For industrial countries, the situation is reversed. The United States has advocated market-based solutions, and the European Community might be following suit. By combining the two distinct elements, the "two sides of the coin," namely, markets and political participation, the IBES could offer a solution that meets the objectives of the two groups of countries.

The following sections explain the background of the climate negotiations in which the proposal of the chapter emerges and how the IBES can help meet the needs of the various nations in the negotiations.

---

[6] As the Federal Reserve does in the United States.

[7] As done by the Federal Communications Commission in Washington, D.C., with the help of auctions of the airwaves.

11.2 Background of the Climate Negotiations: Rio, Berlin, Geneva, Kyoto, and Buenos Aires

11.2.1 The Global Environment — The 1992 Earth Summit in Rio de Janeiro emerged from widespread concern with ozone depletion, biodiversity destruction, and global climate change. One hundred nations met at the Earth Summit to consider reducing the threat of global warming by rolling back emissions of greenhouse gases in the industrialized countries to 1990 levels by the year 2000. The summit emphasized the importance of achieving sustainable development. For this purpose UN Agenda 21, adopted in 1992 by 150 nations, has, as an explicit objective to achieve patterns of consumption oriented toward the satisfaction of *basic needs.*[8]

Despite the interest generated by the Rio Summit, the implementation of its goals has been slow. Part of the problem is scientific uncertainty about the impact of greenhouse gases on the atmosphere. However, science increasingly supports the view that human activity is causing climate change;[9] therefore, this justification for inertia is being removed.

A second and more difficult factor hindering the negotiations is the divergence in the perception of the problem in industrialized and developing countries. Most carbon emissions have originated, and continue to originate, from the industrial countries.[10] Many developing countries take the position that only changes to this pattern can have an impact on the problem, whereas many industrialized countries see the biggest threat in the harm that developing countries can do in the future.[11]

11.2.2 Rio Targets and the Berlin Mandate — After Rio the next most important international meeting on climate change was the Berlin Conference of

---

[8]The concept of *development oriented toward the satisfaction of basic needs* was introduced theoretically and developed empirically by the author in 1977 (Chichilnisky [1] and [2]) in the context of studies of sustainable development in five continents. Following this, the Brundlant Report's definition of sustainable development is also anchored to *basic needs:* "sustainable development satisfies the needs of the present without compromising the needs of the future" (Chichilnisky [5], chap. 2, para. 1).

[9]See the report of the 1996 Intergovernmental Panel on Climate Change (IPCC), which states that there is a "discernable" effect of human activity in the concentration of CO_2 in the atmosphere and the world's climate.

[10]For CO_2, the most important greenhouse gas, the breakdown is as follows: 60% to 70% of all emissions originate from industrial nations currently and about 70% historically, even though industrial countries contain about 20% of the world's population.

[11]Indeed, the 60% reduction that scientists believe might be required to have a substantial effect in lowering the risks of climate change can come only from decreasing the industrial nations' emissions. All developing nations together add up to only about 30% of emissions, and therefore nothing within their power could decrease emissions as required.

the Parties. It concluded on April 7, 1995, by adopting a call for action. It found that the Rio articles were not adequate. A mandate adopted in Berlin—the *Berlin Mandate*—required the negotiation of an emissions-reduction protocol to set hard, quantified limitations on the greenhouse gas emissions in 2005, 2010, and 2020. Another major decision in Berlin was to establish a pilot phase for *joint implementation,* a way in which two nations can cooperate in achieving a reduction in emissions.[12]

Many developing countries have seen joint implementation as a mechanism for transferring responsibility for emissions reduction away from the countries that account for most of the emissions of the planet: the industrialized countries. In addition, because joint implementation is a bilateral process, it can miss many of the opportunities available in multilateral markets and could lead a powerful industrial nation to take advantage of a smaller and less powerful nation in the terms of trade, missing the desirable equal treatment that prevails in competitive markets. To address these concerns the FCCC decided that industrialized countries may not take credit for any reduction of their emissions during the pilot phase, toward their commitments at this stage of the negotiation to reach 1990-level emissions reduction by 2000.

11.2.3 Geneva, Kyoto, and Buenos Aires — Following Berlin, COP2 of the FCCC met in Geneva in July 1996. In the meeting the United States adopted a new position that supports for the first time the concerns of developing countries to establish hard targets for the greenhouse gas emissions by industrialized nations.

Taking a leading position, the Hon. Timothy Wirth, then undersecretary of global affairs of the United States,[13] advocated a market approach for the trading of rights to emit greenhouse gases among the industrialized nations—the approach originally proposed by this author to the UNFCCC earlier in 1994 and presented officially at the third annual World Bank Conference on Effective Financing for Sustainable Development in Washington, D.C., in October 1995. The United States' approach did not, however, go as far as recommending the creation of an international bank for environmental settlements (IBES), which is the natural next step, as argued here.

In December 1997 the COP3 in Kyoto took matters a great deal further. It reached for the first time an agreement for hard quotas from industrial nations[14]

[12] *Joint implementation* refers to one or more parties taking actions or financial actions in the territory of other parties, and it is seen as a prelude to emissions trading by a number of governments and observers.

[13] Currently president of the UN Foundation.

[14] Formally Annex B nations.

by which they will decrease their emissions by 5.2% by the period 2008–12, using as a baseline the level of emissions prevailing in 1990. In addition, the "Kyoto surprise" was an agreement memorialized in Article 12 for the creation of a Clean Development Mechanism (CDM), which incorporates explicitly both industrial and developing nations in a flexible way in the achievement of the Convention's goals. This agreement, still in an embryonic form, is reminiscent of a joint implementation provision that for the first time incorporates both groups of countries. The CDM emerged historically from a proposal advanced by Brazil suggesting the creation of a global development fund that would be capitalized by funds arising from the collection of fees applied to nonperforming abatement duties by developing countries in the context of this protocol. Finally, the Kyoto Protocol introduced Article 17, which is an embryonic agreement on the creation of emissions markets among the industrial nations.

In summary, the Kyoto Protocol limits industrial nations' emissions and provides three "flexibility" mechanisms to help achieve these limits: joint implementation, the CDM, and emissions trading. Of these, only the CDM incorporates both the industrial and the developing nations. The three mechanisms could, however, be linked in the future in innovative ways, together with the technology transfer issue that is crucial for breaking the link between carbon emissions and economic progress.[15] The financial mechanisms implicit in such linkages would be the natural sphere of the IBES. Indeed, Article 12 provides for the creation of an executive committee to monitor the execution of the CDM, which could be a natural overseeing body for the activities associated with the IBES. Following Kyoto, COP4 took place in Buenos Aires in 1998 in order to start the process of developing, completing, and refining what was achieved in Kyoto. Much work remains to be done, and despite the successful advances at Kyoto, the road ahead seems steep and hard.

A brief summary of the political issues involved in the climate negotiations might help explain where the main roadblocks are and how the proposals advanced here could help meet the concerns of the various parties.

11.2.4 Key North-South Issues in the Global Negotiations — Developing countries fear the imposition of limits to their growth in the form of emissions restrictions, on the use of their own resources, as well as unrealistic population targets. Because most environmental damage currently originates and originated historically in the industrialized countries, whose patterns of develop-

[15]Such connections were proposed by the author at the workshop "From Kyoto to Buenos Aires: Technology Transfer and Emissions Trading" with the participation of the major players in the global negotiations at the Italian Academy of Columbia University in April 1998.

ment are at the root of the environmental dilemmas we face today [9, 15], the developing countries have consistently required that the industrial countries take the lead in reducing emissions. To a certain extent the Kyoto Protocol has met this requirement, as its emission limits are placed solely on Annex B nations.

In the Buenos Aires COP4, November 2 to 14, 1998, China, India, and the OPEC countries played an important role in holding up the position of the developing nations. Within the group of developing nations, the members of the OPEC are especially concerned with the changes that the protocol decisions could precipitate in their export markets if petroleum prices increase. A similar position is taken by other resources-intensive exporters, such as Australia.[16] The island nations (Bangladesh, Indonesia, Marshall Islands, and Maldives) are an especially vulnerable group whose plight represents a challenge to humankind. Nothing has been done to address their concerns so far.

Industrialized countries have a different set of concerns. They fear excessive population growth in developing countries and the environmental damage that it could bring. While recognizing their historical responsibility for excessive environmental use, they focus on a long-term future in which global environmental problems could originate mostly in developing countries. The U.S. Senate and the House have voted not to implement any agreement that does not include a commitment on reducing emissions by the developing nations. The United States is the largest emitter (at present about 25% of all carbon emissions originate in the United States), and together with Japan it could block the ratification of the Kyoto Protocol. The North-South issue therefore has practical consequences for the global negotiations. Their success depends on resolving the North-South divide, a divide that has been present since the beginning of the negotiations.

The climate negotiations demonstrate the pivotal role of the developing nations in the process. Indeed the future of industrialization is in the hands of the developing nations. Because industrialization has led to the global environmental problems we have today, if the developing nations were to industrialize and retrace the steps of the industrial nations, the problems' severities would increase severalfold. Simultaneously, the Bretton Woods institutions have traditionally advocated resource-intensive development policies in the developing nations. The traditional style of development based on the intensive and extensive extraction of resources, which are exported and overconsumed in the industrial nations, has come to its logical end. It must be replaced by another

[16]The author gave invited presentations to the Group of 77 and to the OPEC nations providing the recommendations embodied in this article at a workshop organized by UNDP in UN headquarters in New York, September 2 and 3, 1998, and in OPEC headquarters in Vienna, October 28 to 29, respectively.

form of development, the aspiration for which has led to coin the phrases "sustainable development" or "clean industrialization." Here I refer to the knowledge revolution as it evolves and is transmitted throughout the world economy. At the level of the negotiations, however, we are still facing a North-South divide. The road ahead is long and steep. International agreements are customarily adopted by consensus. How to achieve this? It seems that the policies suggested here could set up a cooperative process for industrial and developing nations in the achievement of the goals of the Climate Convention. The following will explain why and how.

11.3 Win-Win Solutions

Implementing the Kyoto Protocol requires a substantial and concerted effort on the part of all parties to communicate and understand each other's concerns, to address in depth the problems and possible solutions, and to reach consensus. An updated understanding (developed here) of the economic aspects of the issues is valuable because it can foster that consensus. In developing consensus it helps to build from common interests. Whereas the main concerns are ecological and environmental, the main stumbling blocks in reaching solutions are economic. To abate carbon emissions means, in the short term, burning less fossil fuel and therefore producing less energy. This means less economics output. This leads to a natural question: Who should abate?[17]

Both industrialized and developing countries face significant abatement costs in the short run because current patterns of development are resource intensive and it is costly to change them. Although the outcome of the policy is uncertain because we know relatively little about the impact of human activity on the environment of the planet, the risks we face are nevertheless sufficient to make it compelling that precautionary steps be taken now.[18] How much is it worth paying to improve our environment, and who should pay? Here I discuss who should abate and why, the role of public goods in determining the outcome, and how to arrive at a cooperative solution that can help bring about consensus.

11.4 The Economics of Climate Change: How to Determine Emissions Limits?

A range of policies to limit emissions trading have been discussed in chapter 2 of this volume. This covers command-and-control instruments that establish

---

[17] See Chichilnisky and Heal [10].

[18] See Chichilnisky and Heal [9].

bounds on economic behavior, taxes, joint implementation, and markets for emissions permits. Chapter 2 also explains how markets with emissions trading work. The simplest form of emissions markets restrict total quantities emitted (as done by the Kyoto Protocol for Annex B countries) and allows countries the freedom to make choices about how to implement these limits and within these limits to trade quotas among themselves (Article 17). A country will buy permits if it wants to emit more than its quota and will sell them otherwise. Prices are flexible, determined by supply and demand.

As already pointed out, the implementation of the Rio targets and the Kyoto Protocol require a measure of consensus about the policy instruments to be used. These policy instruments are new: Emissions trading involves the trading of commitments to reduce emissions, which can be understood as trading "temporary rights" to emit. These instruments share a novel and unusual characteristic. Rights to use the atmosphere of the planet to emit CO_2 are rights to use a *public good:* the planet's atmosphere. As explained below, this unusual characteristic means that unaided markets to trade emissions permits cannot reach efficiency solutions and that backup institutions are needed for the trade in public goods.

Another new aspect of the environmental problem is that emissions, although producing a public good in the quality of the atmosphere, are not produced by governments as are the standard public goods such as law and order.[19] Every person on the planet emits greenhouse gases through driving a car, heating their homes, or producing energy by burning fossil fuels. Emissions markets are therefore markets to trade *privately produced public goods.* Such markets are quite different from standard markets. The allocation of rights to use privately produced public goods requires special attention.

Although the Kyoto Protocol has reached an agreement on limiting industrial countries' emissions such an agreement seems difficult to achieve with the developing nations without first reaching an understanding of what would be fair and efficient at the global level. The Kyoto Protocol limits the emission of Annex B countries, requiring a 5.2% reduction in their emissions by the period 2008–12. Under current patterns this means a reduction of about 30% from projected emissions. If developing nations would join this part of the protocol, how should their emission limits be decided? The general question is, Who should contribute most of the improvement of the atmosphere, to the recovery of the "global commons"? One answer often heard is that this should be the developing countries because they have lower abatement costs. This answer is

---

[19]In contrast with the classic case examined by Lindahl, Bowen, and Samuelson, the public good that interests us here is *privately produced.*

based on the belief that abatement of carbon emissions costs less in developing countries and that abatement carried out in developing countries would achieve the same goal in lower dollar terms and ensure efficiency. Is this argument valid? Only in markets with private goods. In markets for public goods, it is not the dollar value of the abatement that counts for efficiency but rather the opportunity cost of that dollar value in terms of the utility that it can provide. The point is that the same dollar provided brings about very different utility gains in a rich country than in a poor country. Marginal utility gains are what counts to determine efficiency. Chapter 7 in this volume establishes the point rigorously. Here I provide a simple example.

Suppose that abatement of an extra cost ton of carbon costs $1.00 of output in India and $2.00 in the United States. Abatement of an extra ton of carbon costs less in India. Who should abate? The real loss of utility from abatement in India can be much higher than in the United States because $1.00 of goods can have a major impact on the average citizen of India, whereas a $2.00 loss in the United States has only a marginal impact for the average citizen. The point is simple: The marginal utility of income decreases with income. The more income we have, the less our utility increases with the additional dollar. There is a separate but parallel argument from the supply side: Each dollar invested in developing nations leads on average to more production than a dollar invested in industrial countries (see the data in the Appendix) so that if abatement reduces investment, initially it should take place in industrial nations, for efficiency.

These matters do not count in economies with private goods because everyone chooses independently of one another and traders can adjust their consumption to equate the marginal gains they derive from the markets.[20] However, with privately produced public goods they do. In these cases the condition of equal marginal costs is not appropriate for efficiency.[21] It is appropriate only when all countries have the same marginal utility of income. In other words, only when (free) transfers are made between countries so as to equate their marginal valuations of private consumption does efficiency require that marginal abatement cost be equal. However, such transfers would be unrealistically large.[22] Therefore, in general, efficiency implies that abatement will come pro-

[20]Marginal rates of substitution must all be equal across markets and must equal the marginal rates of transformation in those markets.

[21]See Atkinson and Stiglitz in reference [1] of chapter 1 in this volume. The rule is typically that the sum of marginal rates of substitution equals the marginal rate of transformation when the government produces the public good. See Chichilnisky, Heal, and Starrett [13] for the case in which free international trade in permits is allowed. The answer is the same.

[22]Paid transfers, such as those that occur within international markets, need not equate the marginal utility of consumption across trading regions.

portionately more from those countries that have higher income because they have a lower marginal utility from increased consumption than poorer countries.[23] Under general conditions, the proportion of income dedicated to abatement should increase with the level of income. Therefore, an answer to the question "Who should abate?" should be: First of all, the industrialized countries.[24] This has been the position of the developing countries for many years. As reported previously, even the United States agreed with this position in Geneva in June 1996, and now the Kyoto Protocol signed in November 1998 by the United States makes implicitly the same point. The Kyoto Protocol provides only for abatement obligations on the part of the industrial nations.

Requiring abatement from developing countries first would be a regressive measure, like taxing the poor the most. There are other concerns about regressive measures. They can cause problems because environmental degradation and poverty are closely connected. Anything that worsens poverty is likely to lead to further environmental degradation and to increased rates of population growth.[25] For example, a policy that lowers the price of wood and therefore the income of harvesters can lead to more than less extraction of wood [16]. Because the purpose of taxing the price of wood is to discourage extraction of wood, by decreasing the income of the harvesters the tax could achieve the opposite effect from what is intended.[26]

Until now the issue of user's rights on the atmosphere has been left to the political arena, with the understanding that it involves exclusively a transfer of wealth between countries. An implicit assumption is that markets themselves function efficiently; the matter to be decided was the distribution. The two issues, efficiency and distribution, were seen as separate. The latter, distribution, was seen as a major political hurdle and a divisive issue that complicated matters and interfered with the development of consensus. Emissions trading has as its goal an efficient allocation of emissions within the global limit. However, in order to trade, one must know who owns what. This means that users' rights must be established: One must establish who has the rights to emit and how much. This is not necessary for taxes, but it is for markets.

Building on recent advances in the economics of climate change presented elsewhere in this book, the Appendix shows a somewhat unexpected source of common interest among industrialized and developing countries.[27] There is

---

[23] See Chichilnisky and Heal [10].

[24] See Chichilnisky and Heal [10].

[25] See, e.g., *World Development Report* [15], 1992.

[26] See Chichilnisky [16].

[27] See the Appendix, Chichilnisky [19], Chichilnisky and Heal [10], and Chichilnisky, Heal, and Starrett [13] and chapter 3.

a new role for distributional issues: The appropriate equitable distribution is needed for markets to function efficiently. Somewhat surprisingly, a measure of equity can lead to efficient allocation.

11.5 Win-Win Solutions in the Climate Negotiations

The somewhat unexpected link between the distribution of emission limits and overall efficiency established in the CHS chapter (3) and extended here presents an opportunity for advancing the climate negotiations: a source of common interest between industrialized and developing countries. Efficiency is often favored by industrial countries that have the most developed markets, whereas equity is an issue that concerns the developing countries most. It seem useful to explain intuitively how the connection between equity and efficiency arises in this context; for the formal results the reader is referred to prior work[28] and to chapters 2 and 3 and the Appendix.

Efficiency in a competitive market requires that the total amount emitted across the globe, which determines the quality of the atmosphere for all, be precisely the choice that individual traders themselves would make independently, given their other holding of private goods. The connection between distribution and efficient operation of the world economy stands in sharp contrast with the properties of markets for private goods. With private goods, no matter what the distribution of property rights, an efficient allocation is always reached by a competitive market. When markets trade private and public goods simultaneously, they achieve efficiency only when the initial conditions are such that the traders who own fewer private goods own more users' rights on the environment than the rest. Market efficiency requires a somewhat flexible but inverse relationship between property rights in private goods and property rights in public goods. In practice this means that industrialized countries, which have a much larger initial allocation of property rights on private goods, should initially be given relatively smaller endowments of property rights on public goods as a precondition for market efficiency. This unique property of markets with privately produced public goods is developed formally in the Appendix and leads us to the policy proposal of this chapter: the creation of an IBES.

11.6 IBES: A Self-Funding Mechanism?

In contemplating a new global financial institution, a natural question is how to fund it. The Bretton Woods institutions are funded by voluntary contribu-

[28] See Chichilnisky [19], Chichilnisky and Heal [10], and Chichilnisky, Heal, and Starrett [13].

tions from the rich countries that are collected from taxes raised in their territory. However, voluntary contributions have declined and seem more difficult to achieve in today's political climate; the continuing and escalating indebtedness of the United States with respect to its dues to the United Nations offers a good example. Using the same voluntary approach the Global Environment Facility (GEF) requires periodic replenishments of its fund in a difficult environment in which aid has fallen well below the amounts targeted by the United Nations.[29] Funding a new institution using existing voluntary mechanisms seems therefore unrealistic.

The recommendation I have proposed is the creation of global financial mechanisms that are self-financing. This might be possible in some cases and not in others; for example, humanitarian disaster aid cannot generally be self-financing, and it would be counterproductive to ignore this fact. However, in the environmental area several possibilities exist for self-funding mechanisms, for example, by developing financial instruments that use *as collateral the environmental assets of the planet.* This possibility emerges from the provisions of the Kyoto Protocol, which can be a basis for developing self-financing mechanisms that do not rely on taxation or voluntary contributions. Indeed the limits on emissions that it sets for Annex B countries create de facto a new store of economic value arising from the scarcity in the use of the atmosphere implied by the Kyoto Protocol's emission restrictions. Limits on emissions and the ability of trading unused credits as provided in Article 17 create a source of value that can be realized in environmental markets.

The type of institutions that we have in mind is crystallized in the IBES,[30] but the type of solutions can take many forms and are not restricted to the creation of a single institution. Global environmental assets include the world's forests and bodies of water, its minerals, and biodiversity. These include some of the most valuable resources known to humankind, on which depends our ability to survive. Yet today most forests in developing nations (such as Ecuador and Brazil) are destroyed to produce minerals and agricultural products for sale in the international market. The right financial mechanisms are needed to realize their value without destroying them. An analogy is provided by traditional mortgages, in which assets (such as buildings) serve as collateral for obtaining financial value from the asset (the building) without destroying the asset itself. Without mortgages the only way to obtain value from a building might be to break its walls and sell the bricks one by one in the market. This is

---

[29] Overseas development assistance (ODA) was targeted at 0.7% gross domestic product (GDP) of the industrial countries, but it is close to one-third of that target at present.

[30] Chichilnisky [14].

possible, but it is not economically desirable: Little money would be obtained, as the value of the building is much larger than the sum of its bricks, and the building itself would be destroyed in the process. Today's economic policies toward the environment have a similar flavor. Often we destroy enormous and valuable ecosystems by selling their trees one by one because the economic need is pressing and in many cases because no one has a clear title to the property, so that they treat it on a first-come, first-served basis. This situation is typical in developing nations that hold resources under a common property regime and leads to overexploitation of resources that are exported to industrial nations at prices that are below replacement costs (Chichilnisky [16]). Resolving this situation might require institutional arrangements for clarifying, assigning, and protecting property rights when needed and organizing, executing, and monitoring the trading of emissions permits, loans on these, and derivative instruments associated with them. The sections below show (1) how an IBES could work in practice, and (2) why the role of such a global institution would complement markets but go much further than anything that unaided markets can achieve.

11.7 How the IBES Would Work

The IBES would be led by industrial and developing nations, represented politically in an equal footing, extending the current negotiating role of the FCCC to a continuing management role on behalf of the international community. The IBES could provide the backbone of the global environmental markets, extending existing institutions to the global level and ensuring their efficiency and integrity.

Markets involving SO_2 nitrous oxides (NOX) and various water pollutants constitute interesting precedents for the IBES. In 1993 the Chicago Board of Trade introduced SO_2 emissions trading, following the United States Clean Air Act, which introduced ceilings and rights to emit for US utilities. These markets are regulated by the Commodity Futures Trading Commission to ensure their efficiency and integrity. The SO_2 markets are less appropriate than CO_2 markets because, as opposed to CO_2, SO_2 does not mix uniformly and stably in the atmosphere. As a result, trading between states can lead to violations of the Clean Air Act because states that buy more permits can end up with higher emissions levels. In addition, the primary traders are rather few, therefore offering little market depth: about 150 utilities in the United States as a whole. Since localized trading is necessary in some cases this additionally limits market depth. However, SO_2 markets have been rather successful in helping implement the ceilings of the Clean Air Act at relatively little cost, leading to about $14 per ton of emissions saved. Similarly, the IBES could also do the following:

1. fulfill the role of a clearing and settlement institution
2. offer credit enhancements for the carbon emissions permits sold by adding credit worthiness to contracts and perhaps by ensuring that the counterpart to each contract is the bank rather than another country or corporation, as in the case in the commodities clearinghouse
3. determine which type of instruments will be traded—for example, derivative securities (options or futures)—and if so, how
4. serve as a forum for recording environmental accounts that could be used to monitor the successes and failures of implementation
5. regulate the relationship between primary and secondary markets, a matter of great importance in ensuring market liquidity
6. run open-market operations and, in general, have an impact on borrowing and lending rates, such as the Federal Reserve does in the United States and all central banks do around the world

In addition to CO_2, other environmental markets could be involved in the IBES, such as water markets and markets for biodiversity use. The IBES could incorporate other environmental markets and financial mechanisms: water markets, such as those currently emerging in southern California, and markets for trading environmental risks, such as hurricanes, which are believed to have become more unpredictable and violent owing to the global climate change.[31] Chichilnisky and Heal studied the securitization of watersheds. Recently, a proposal for securitizing the emission reductions attendant to clean technology transfer on the basis of emissions markets was advanced in Chichilnisky [18].

11.8 The Role of the Regional Banks

As part of the FCCC system, the IBES could offer developing nations the ability to participate in orderly voting procedures to regulate and monitor the performance of the global emissions markets, the periodic allocation of emis-

[31] In 1992 the creation of an instrument that would offer contracts contingent on an unknown frequency of losses was proposed (in Chichilnisky and Heal, [9]) that is now traded on the Chicago Board of Trade under the name of Catastrophe (CAT) Futures. Another instrument has been proposed more recently, obtained by "bundling up" mutual insurance contracts as well as securities. Chichilnisky [17] studies the use of profit-sharing agreements to obtain value from biodiversity without destroying it, using the Merck-INBIO deal as an example, and proposes deeper access to capital by securitizing such deals.

sion reduction obligations by the different countries and regions, and the monitoring of the compliance with the contracts. To achieve consensus on the voting rules within the IBES, these could be adapted from existing regional banks' procedures, involving participants from industrial and developing nations, such as the Interamerican Development Bank (IADP), in which 50% of the vote is in the hands of the borrowers and 50% in the hands of the lenders. In addition to the IADB, other regional institutions such as the Asian Development Bank and the African Development Bank could participate in creating a task force of the world's regional development banks that would be in charge of creating and offering credit enhancement for the securities that IBES would offer. This would attract private financing for clean technology products in the various regions. These securities could be backed by certain assets: the emission reduction certificates corresponding to each clean technology project. Once these certificates are traded in the Annex B market provided by Article 17 of the Kyoto Protocol, they acquire a market value; however, this value is in the future. Credit enhancement facilities from regional banks would reduce these instrument's risks and therefore make them easier to place in the world's capital markets. In summary, the role of the regional banks would be to help the transition between the present and the future by offering credit enhancement facilities for these securities so that they can be placed in the world's capital markets.

11.9 The IBES Mandate

As part of its mandate, the IBES would ensure the following:

1. The trading of greenhouse gas emissions should not compromise the future ability of developing countries to grow.
2. The trading of emissions rights should not conflict with humanitarian aid or other international flows, such as overseas development assistance.
3. The IBES should provide more access to capital for development. It should not induce selling of emissions rights under unfavorable prices.
4. The trading of emissions rights will be initially among industrial nations. Indeed in the Kyoto Protocol trade is contemplated only among Annex B countries.
5. The IBES should help ensure fair markets and equal access to information and to trading; it will also ensure market integrity and depth.
6. Deals should be structured so that they can be reversed without undue penalty to the selling countries, which may revise their priorities in the future.

11.10 Policy Recommendations

The following policy recommendations have been discussed with members of the FCCC and government agencies various nations:[32]

- *Recommendation 1.* A migration from "joint implementation" to multilateral procedures involving global markets for emissions rights. The emissions markets would involve only industrialized countries initially. The recommendation was supported by the United States in Geneva in June 1996.
- *Recommendation 2.* Emissions rights could be loaned, instead of (or in addition to) sold, with the lending and borrowing managed by the IBES.[33] A key aspect of a loan rather than an outright sale of emissions rights is that developing countries need not be concerned about unforeseen long-term consequences of an irreversible transfer of their emissions rights to other countries or with making irreversible deals at prices that will subsequently look unreasonable. Lending rather than selling these rights avoids many uncertainties faced by developing countries entering into an emissions abatement agreement. Furthermore, lending rates can be regulated by the IBES.
- *Recommendation 3.* Developing countries may wish to lend emissions rights for limited periods until their needs for these are clear, whereas industrialized countries are likely to want to borrow for longer periods. The IBES could match these positions by borrowing short and lending long in the traditional manner of financial intermediaries. In exchange for the risk involved, it would charge a borrow-lend spread. Commercial capital and international financial institutions, private or not, would undoubtedly be attracted to such as operation.
- *Recommendation 4.* To ensure fair prices to developing countries, it might be desirable for the IBES to establish a market rate of interest on emissions permits in a market open only to industrialized countries and then to pay this rate on deposits from developing countries.

---

[32]Including Minister Raúl Estrada-Oyuela, 1994 chair of the INC/FCCC and 1997 chair of the Negotiating Committee of the Kyoto Protocol; H. E. Ismail Razali, ambassador, permanent mission of the Malaysian to the United Nations; Mr. Xialong Wang, third secretary, Chinese permanent mission to the United Nations; Mr. James Baba, deputy permanent representative of Uganda to the United Nations; Dr. John Ashe, counsellor, permanent mission of Antigua and Barbuda to the United Nations; and Carlos Sersale de Serisano, currently special adviser to the secretary general of the UN Industrial Development Organization (UNIDO).

[33]I am grateful to Geoffrey Heal for this suggestion.

- *Recommendation 5.* It might be desirable to securitize carbon emission reductions from new technologies and products so as to attract funding from global capital markets and generate self-funding mechanisms to fund such technologies and products, therefore fostering clean industrialization and advancing the knowledge revolution in developing nations.[34]
- *Recommendation 6.* It might be desirable to securitize some of the world's watersheds in order to attract private funding for the conservation of clean water resources.[35]
- *Recommendation 7.* The securitization of the planet's biodiversity and global reinsurance of environmental risks that are associated with developing areas could be equally handled by the IBES.
- *Recommendation 8.* A similar treatment of the earth's airways would be desirable.
- *Recommendation 9.* The establishment of a system to monitor and account for the successes and the failures of the trading agreements should be developed.

11.11 Conclusion: Resource-Intensive versus Knowledge-Intensive Growth

This chapter argues that new institutional mechanisms are needed to achieve the goals of the Climate Convention and implement the Kyoto Protocol, especially with respect to the emissions markets provided for in its Article 17. This is because of the idiosyncratic nature of these markets, which require special patterns of users' rights, favoring lower-income groups, in order to achieve efficient use of resources. An institution, the IBES, was proposed, and its role was specified as leading the development of the world economy in a new form of clean industrialization the way that the Bretton Woods institutions led the world economy into resource intensive industrialization after World War II. A ray of hope that requires careful consideration is the knowledge revolution, which the IBES could help orient into a resource-conserving direction. The knowledge revolution is a global trend that is taking place whether or not the Climate Convention reaches its objectives. The most dynamic sectors in the world economy today are not resource intensive but knowledge intensive: biotechnology and entertainment, software and hardware, communications, and

[34] See Chichilnisky and Heal [12].

[35] See Chichilnisky and Heal [12].

financial markets. These sectors are relatively friendly to the environment, use relatively few resources, and emit little CO_2; figures 11.1–11.7 illustrate the case of the United States. Knowledge-intensive sectors include financial markets and health services, consumer electronics and telecommunications, and biotechnology. These are the high-growth sectors in the United States and in the most industrialized countries and are developing rapidly in other regions of the world, such as Singapore, parts of India, Bermuda, and Barbados. See the figures provided in the Appendix. Some of the most dynamic developing countries are making a swift transition from traditional societies to knowledge-intensive societies. Mexico produces computer chips, India's Bangalore is fast becoming one of the world's largest exporter of software,[36] and Barbados has recently unveiled a plan to become an information society within a generation. There is nothing new about policies that steer a nation away into knowledge-intensive growth. These are precisely the policies followed by the Asian Tigers: Hong Kong, the Republic of Korea, Singapore, and Taiwan Province of China, all countries that have achieved extraordinarily successful performances over the last 20 years, not relying on resource exports but rather knowledge-intensive products such as consumer electronics. By contrast Africa and Latin America emphasized resource exports and lost ground.

The lessons of history are clear, steering us away from a reliance on resource exports as the foundation of economic development. Africa and Latin America must update their economic focus. Indeed the whole world must shift away from resource-intensive economic processes and products. In doing so fewer minerals and other environmental resources will be extracted, and their price will rise. This is as it should be because today's low resource prices are a symptom of overproduction and inevitably lead to overconsumption.[37] Not surprisingly, from an environmental perspective one arrives at exactly the same conclusion: Higher resource prices are needed to curtail consumption. Producers will sell less but at higher prices.

This is not to say that all will gain in the process. If the world's demand for petroleum drops, petroleum producers might lose unless they have diversified into products that involve fewer resources and higher value. Most international oil companies are investigating this strategy. The main point is that nations do not develop on the basis of resource exports, and at the end of the day development can make all better off. As the trend is inevitable, the sooner one makes the transition, the better.

---

[36] Bangalore exports at present about $2 billion worth, having initiated this sector about 11 years ago.
[37] See Chichilnisky [16].

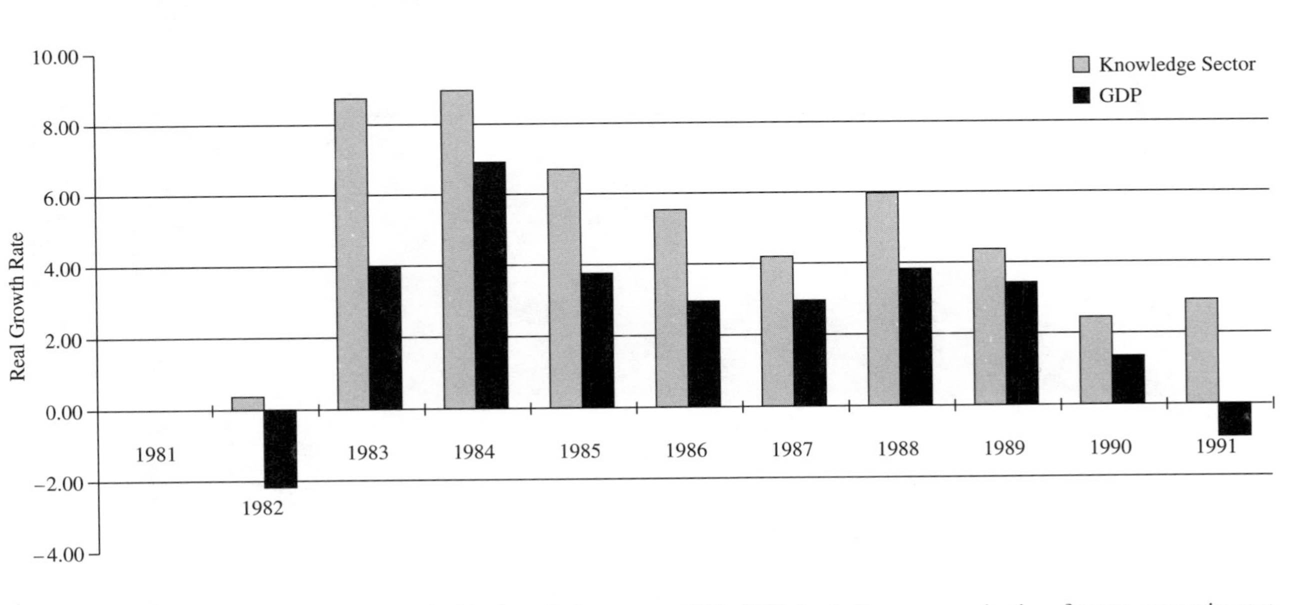

FIGURE 11.1 Growth rate of U.S. GDP and of its knowledge sector, 1982–1991 (including communication, finance, entertainment, electronics, computers and scientific instruments, pharmaceuticals, aerospace, and biotechnology).
Source of figures 11.1–11.7: Chichilnisky and Heal [10].

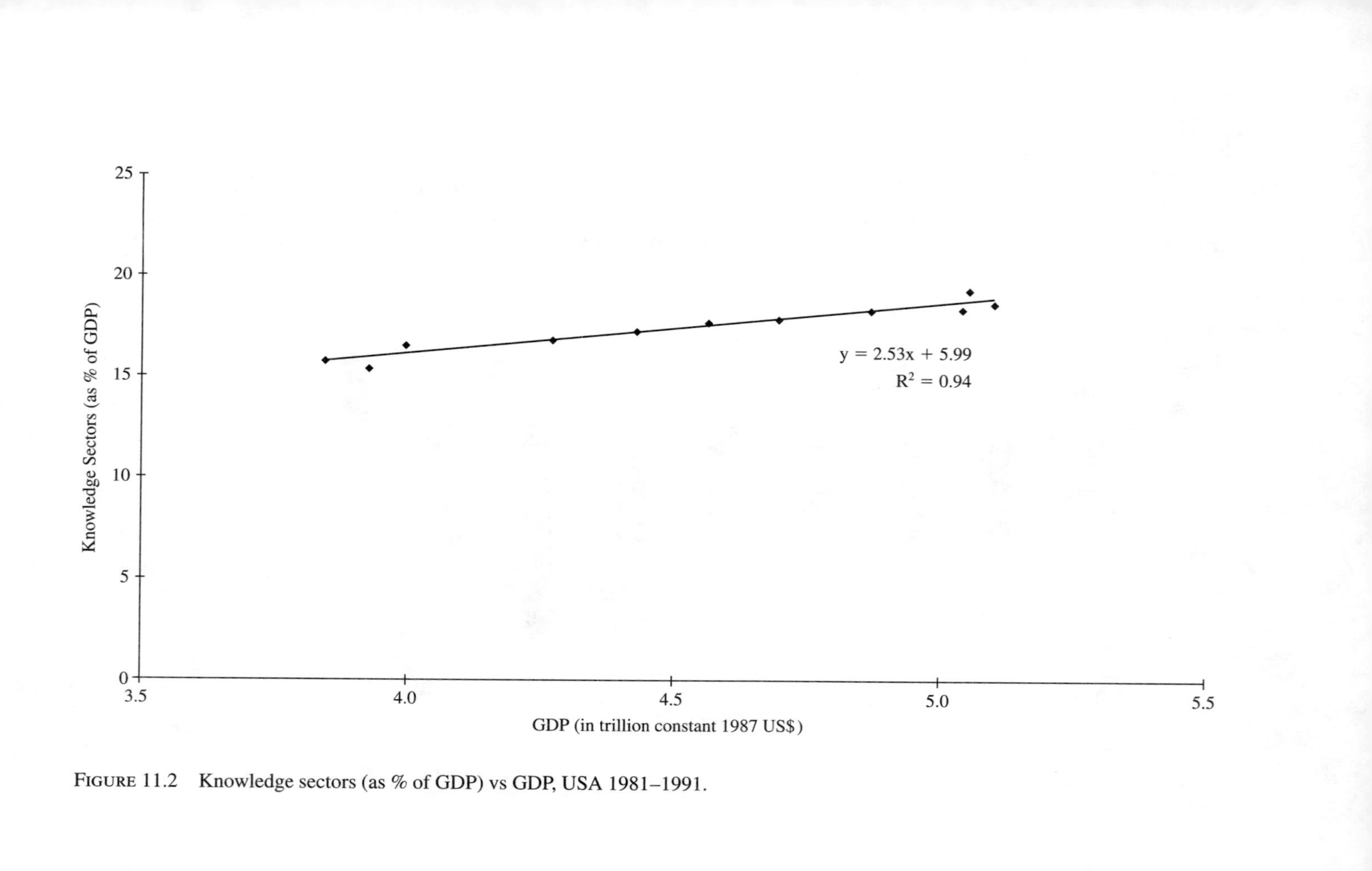

FIGURE 11.2 Knowledge sectors (as % of GDP) vs GDP, USA 1981–1991.

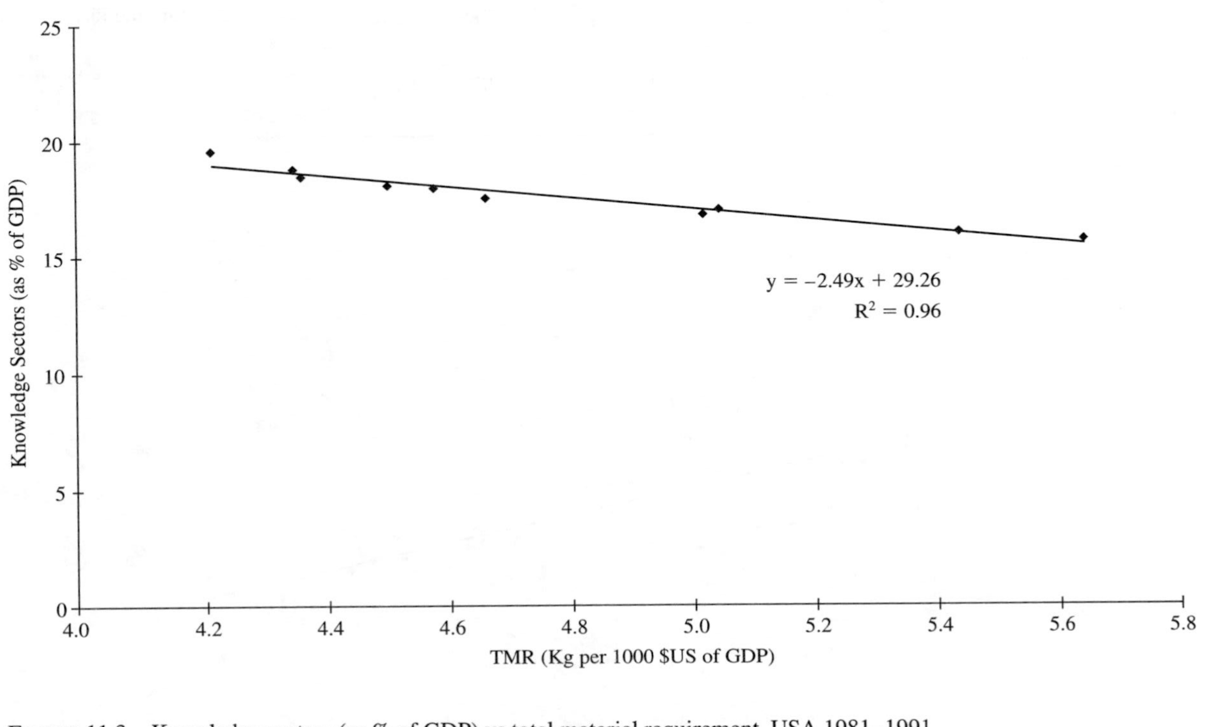

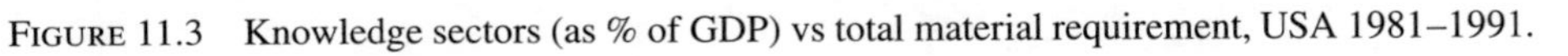
FIGURE 11.3 Knowledge sectors (as % of GDP) vs total material requirement, USA 1981–1991.

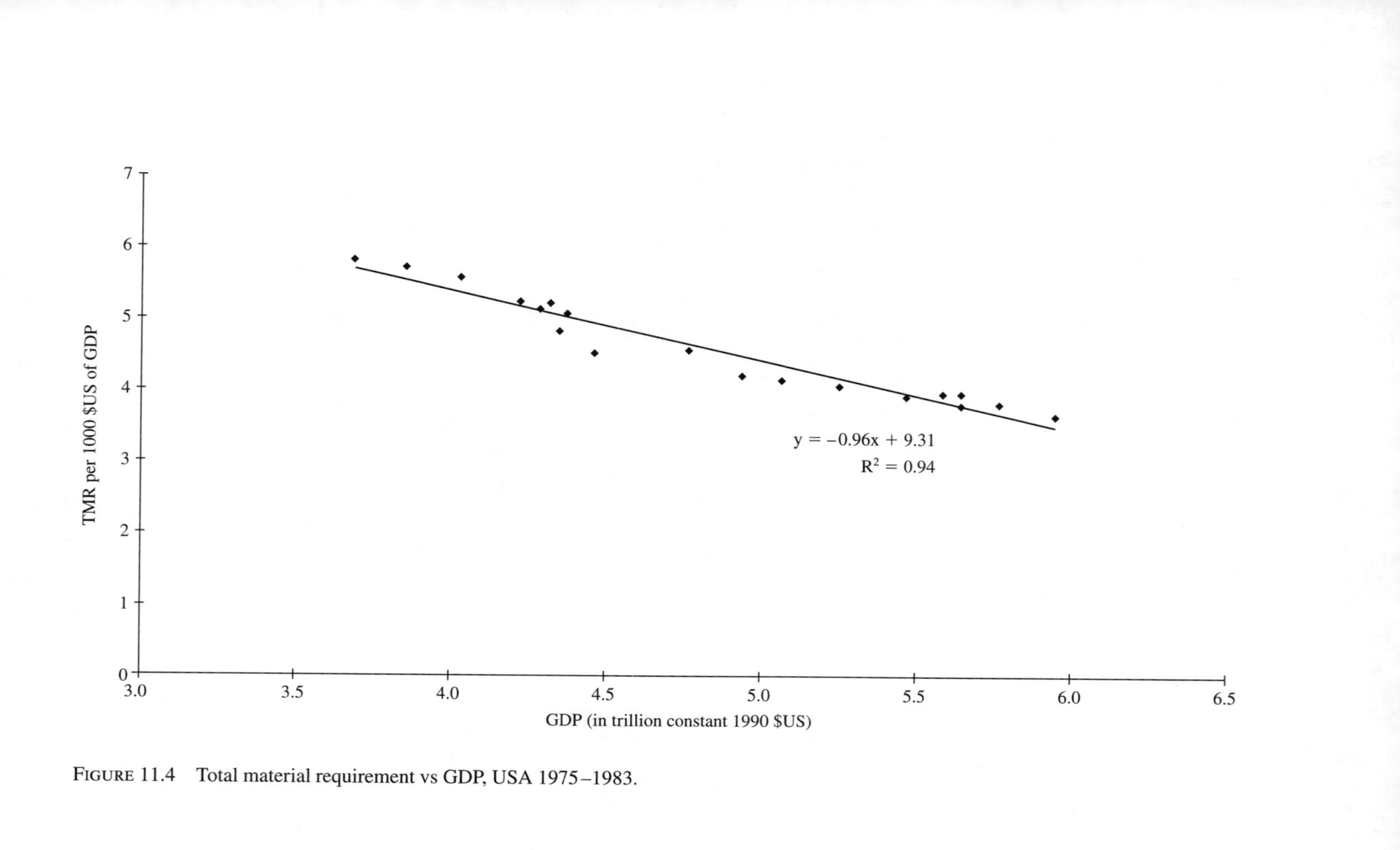

FIGURE 11.4 Total material requirement vs GDP, USA 1975–1983.

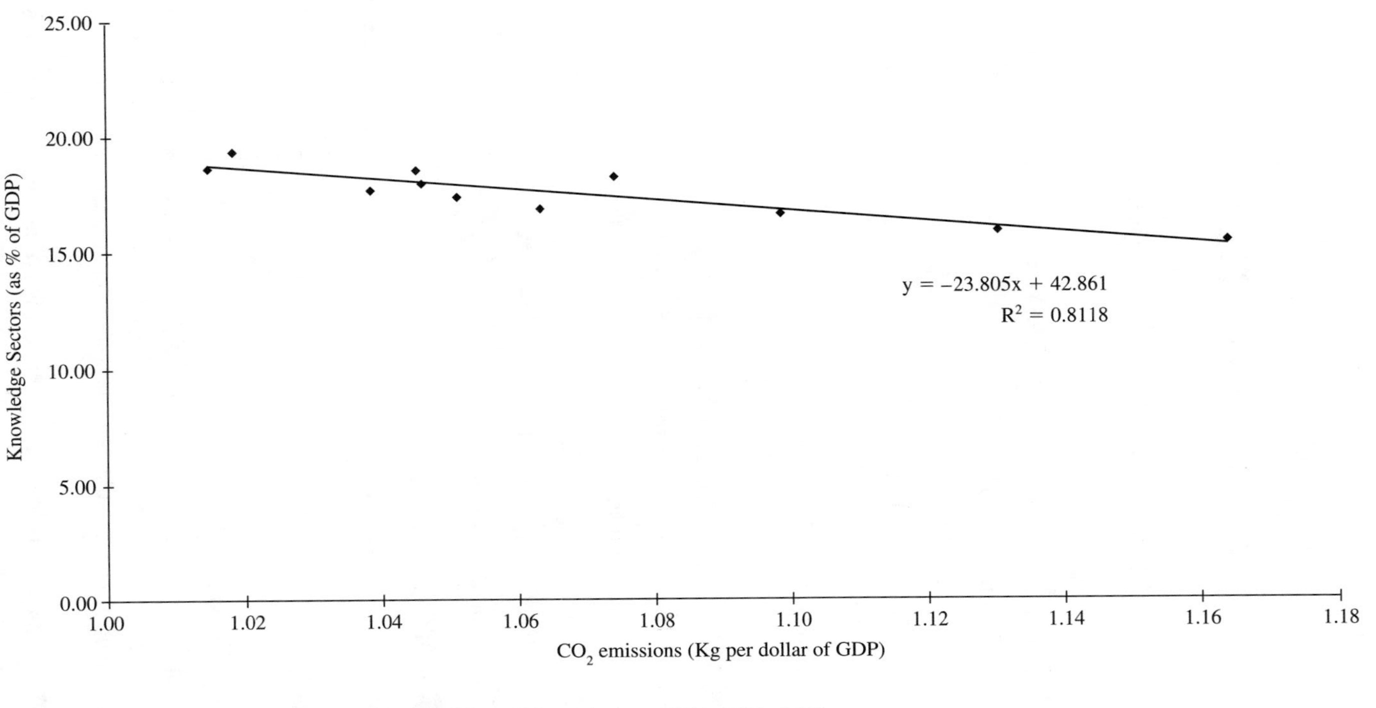

FIGURE 11.5 Knowledge sectors (as % of GDP) vs CO_2 emissions, USA 1981–1991.

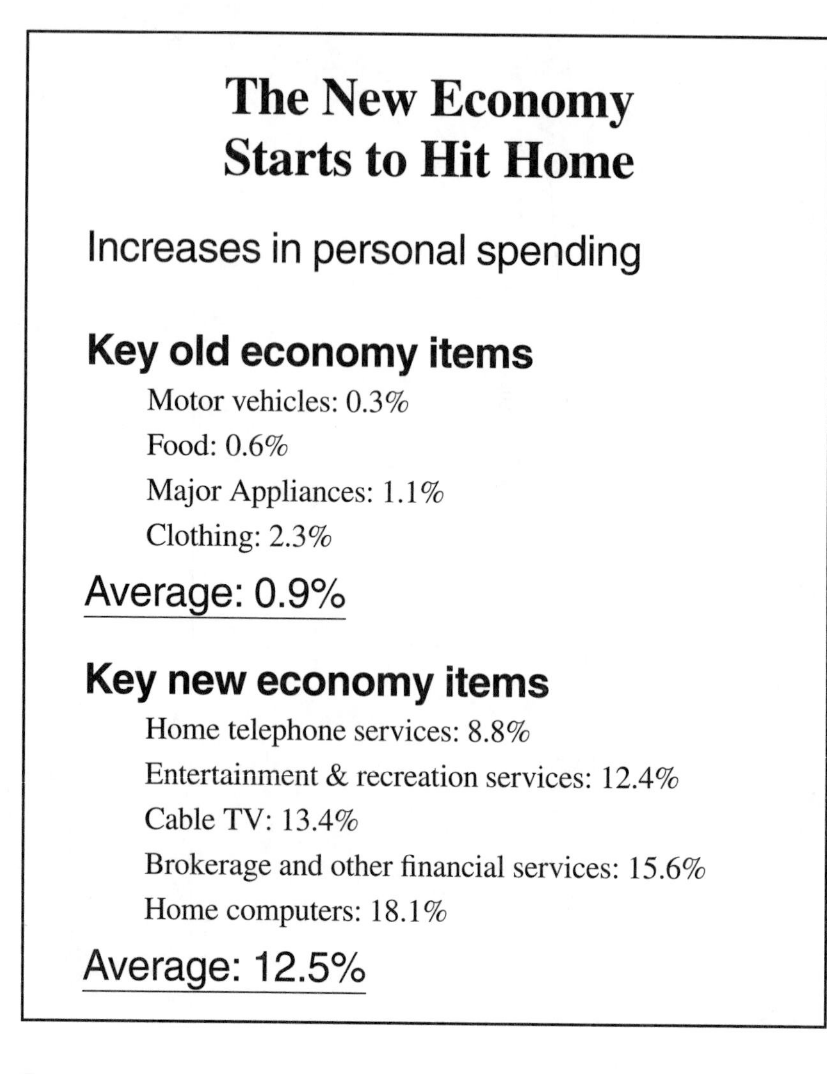

FIGURE 11.6 Increases in personal spending as reported in *Business Week,* March 23, 1998.

To understand the issues and develop policy toward knowledge-intensive development, conceptual advances in economics are needed. The economics of climate change involve challenging questions, such as the following:

1. Which policy instruments or combination of instruments at the national and international levels—carbon taxes, joint implementation, or tradable emissions for CO_2—are preferable for reducing emissions?

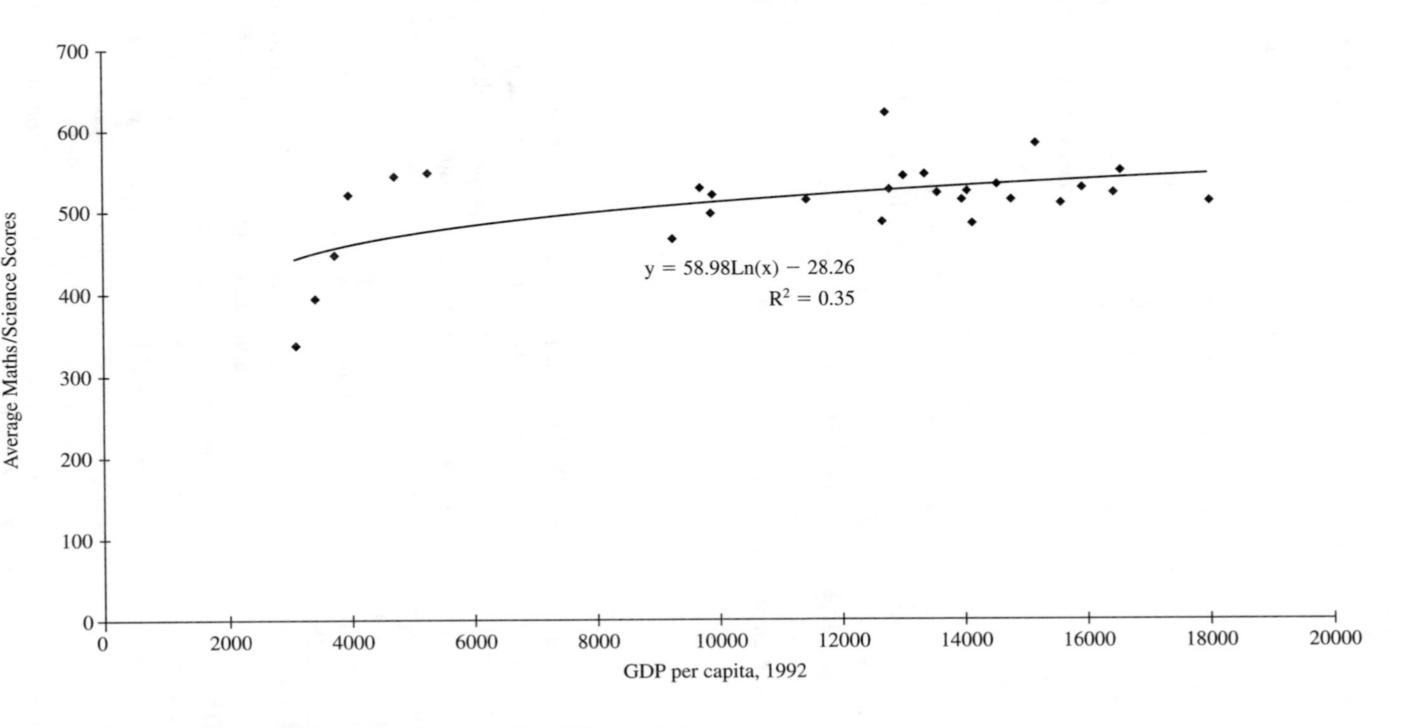

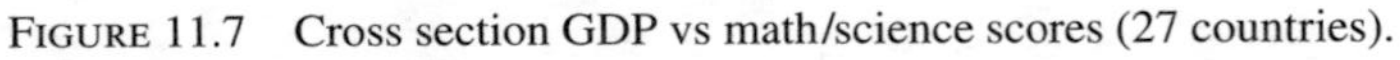
FIGURE 11.7 Cross section GDP vs math/science scores (27 countries).

2. How can an acceptable degree of equity in the use of global carbon be ensured?
3. How would the notions in this chapter impact trade among industrial and developing countries?
4. Which instruments might be needed to support and regulate the trading, clearing and settlement of emission rights and related assets, and to ensure the efficiency and integrity of the market?
5. What type of environmental accounts will help record and monitor the success or failure of taxes, joint implementation schemes, or emissions markets?
6. When do market prices accurately reflect the value of resources, and when should new institutions be created?
7. What is the scope of applying the proposals in this chapter—beyond greenhouse gas emissions—to tackle other cross-border problems, such as desertification or soil erosion and deforestation?
8. What are the implications of the results in this chapter to policies toward markets involving knowledge goods, which, as environmental goods, are often privately produced public goods?

Appendix

A Manifold of Efficient Allocations of Users' Rights

In chapter 3 of this volume, Chichilnisky, Heal, and Starrett (CHS) develop a model of a competitive market with several traders whose utility depends on their consumption of private goods and one privately produced public good, for example, the gaseous composition of the planet's atmosphere. The traders trade private goods as well as the rights to use the public good, for example, the rights to emit. An overall ceiling is placed on their rights to emit that is shared by the traders in fixed given amounts, namely, their respective users' rights (also called "property rights").

This Appendix simplifies the CHS model to one with only two goods and two traders and extends it to allow for a variable limit on total global emissions. This is done in order to show graphically two results that appear in this chapter and not in CHS: (1) With a variable amount of total emissions, the initial allocations of users' rights that yield efficient market solutions define a *one-dimensional manifold of efficient users' rights*. By comparison, in the standard market with private goods, the set of initial allocations that yield efficient market equilibria would be two-dimensional. The implication from this is that efficiency is more difficult to achieve in environmental markets and requires setting up correctly the initial conditions. (2) For each level of global emissions, there is an inverse relation between the initial ownership of private goods and the users' rights on public goods that is needed to achieve efficient market equilibrium.

A brief summary of the model in chapter 3 follows. There are two traders (North and South) trading a private good x and a privately produced public good a, which represents the concentration of greenhouse gases in the world's atmosphere. By definition a is available to both in regions in the same quantity. Each region denoted $i = 1, 2$ produces private goods using as an input different amounts of the public good, that is, emitting different amounts of CO_2: $x_i = \phi_i(a_i)$, $\phi' < 0$. The private good is the numeraire ($p_x = 1$). Trader i has a utility function $u_i(x_i, a)$, which is increasing in both variables, and an initial allocation $\bar{a}_i$ of total amount of emissions, which varies over an open interval $I \subset R$. Total emissions limits are given by $\bar{a} = \bar{a}_1 + \bar{a}_2$, and they vary over the set $I + I \subset R$. For each initial allocation $\bar{a}_1, \bar{a}_2 \in R^2$ of users' rights, a *market equilibrium* is defined by (1) a (relative) trading price paid for the rights to emit, π^*; (2) an amount of the public good used in each region a_i^* to produce private goods (i.e., the emissions) and the amount of emission rights purchased or sold: $a_i - \bar{a}_i$; and (3) an amount of private good produced and consumed by each region x_i^*. In a market equilibrium, each trader maximizes the utility $u_i(x_i, a)$ over the budget set defined by the equation $x_i = \phi(a_i) + \pi(a_i - \bar{a}_i)$, which indicates that the regions' consumption of private goods cannot exceed the value of its production of private goods $\phi(a_i)$ plus the income derived from selling (or buying) permits. In addition, markets clear; that is, $a_1^* + a_2^* = \bar{a}_1 + \bar{a}_2$.

When ϕ and u are smooth, CHS established the following result for a generic set of economies.

THEOREM 1 Given a total level of emissions for the world economy $\bar{a}$, there is a finite way to allocate the rights to emit among the various regions so that the resulting competitive equilibrium is Pareto efficient.

In contrast with the CHS chapter, here the total amount of emissions is allowed to vary; that is, the value $\bar{a}$ is a real variable defined over $I \subset R$; as $\bar{a}$ varies one obtains different equilibria of the world economy. According to CHS, for a fixed $\bar{a}$ the equilibria are locally unique. This follows from Sard's theorem and the global implicit function theorem. In our case, as $\bar{a}$ varies we obtain a larger set of equilibria, and, under generic conditions, this set describes a one-dimensional manifold of the same dimension as the parameter space I. Therefore, for a generic set of two-trader economies as specified previously, Theorem 2 below follows.

THEOREM 2 By allowing total carbon emissions to vary, one obtains a one-dimensional manifold of property rights (rights to emit, or obligations to abate) from which the competitive market with tradable permits achieves a Pareto-efficient allocation of resources in the two-region world economy (Chichilnisky [8]).

PROOF. This follows from the global version of the implicit function theorem, Sard's theorem, and Theorem 1. ■

In a generic two-trader economy as specified previously, one therefore obtains the following:

COROLLARY 3 In an economy with Cobb-Douglas utilities that are the same for all regions, the set of initial allocation of users' rights that lead to an efficient equilibrium allocation exhibits a negative association between the ownership of private and public goods. At the initial conditions leading to the efficient equilibrium, the traders who own smaller endowments of private goods own a higher allocation of public goods and vice versa.

PROOF. See Chichilnisky and Heal [10] and chapter 7 of this volume. ■

Figure 11.8 illustrates how a change in the property rights regimes assigning to the developing nations an increasing amount of rights to emit and fewer rights to emit to the industrial nations can be Pareto improving to all regions. Observe that this result is not possible in markets with private goods in which competitive equilibria are always Pareto efficient.

Simulations on Emissions Trading in GREEN/PIR

Computer simulations were carried out at the Program on Information and Resources (PIR) of Columbia University for the OECD GREEN model modified to incorporate the possibility of trading emissions permits between the countries (hereafter the GREEN/PIR model). This model differs from that of CHS in that there is no environmental quality variable in the utility of the traders: Utility is derived exclusively from the consumption of private goods. Under these conditions the results on equity and efficiency reported previously do not follow, although it is clear that there is no reason to consider abatement of emissions unless there exists a disutility associated with it, making the model less realistic. In any case the runs reported have exhibited a result similar to that discussed previously, although in a different sense. We say that a run is

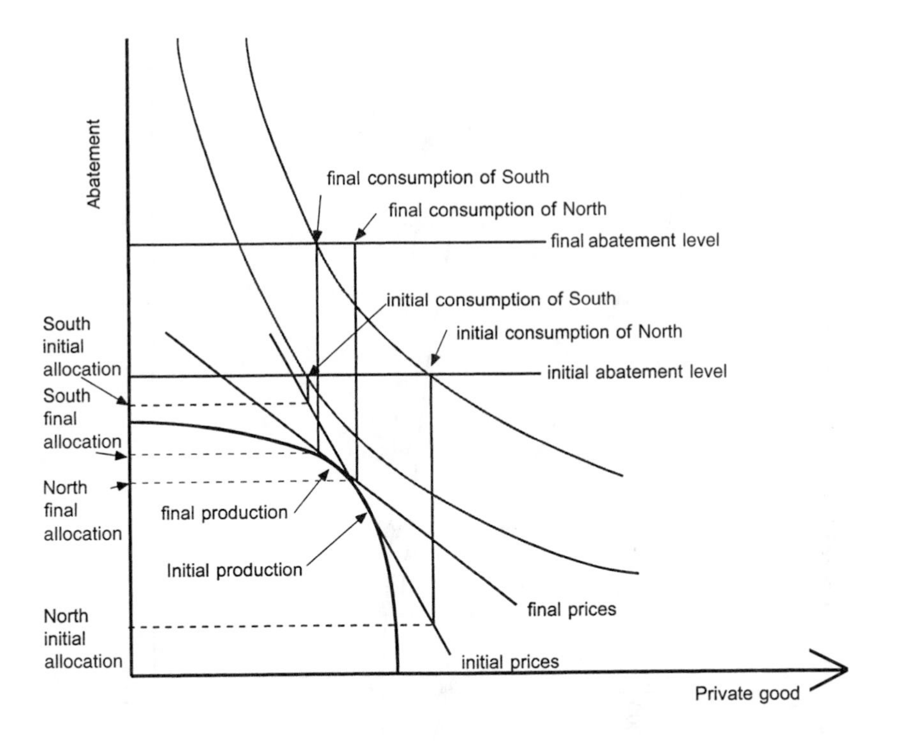

FIGURE 11.8 Only specific distributions of property rights lead to Pareto efficiency.

more efficient than another when it achieves the same level of carbon emission reductions with higher amount of private goods produced. In the runs reported here, it is shown in table 11.1 that the most efficient runs, in terms of minimizing the loss of economic growth that abatement induces, are those in which the distribution of emission permits favors the developing countries. In observing why this happens within the GREEN/PIR model, it appears that the productivity in developing nations (such as China) is on average higher than in industrial nations, so that the abatement of a ton in carbon in industrial nations decreases economic growth by less than it would do in China. Because China imports private goods from the industrial nations, the final result is that all benefit from the abatement rule adopted.

Empirical Analysis

The experience of the last 20 years confirms the GREEN/PIR simulations. On average, a dollar invested in a developing nation has a larger productivity than

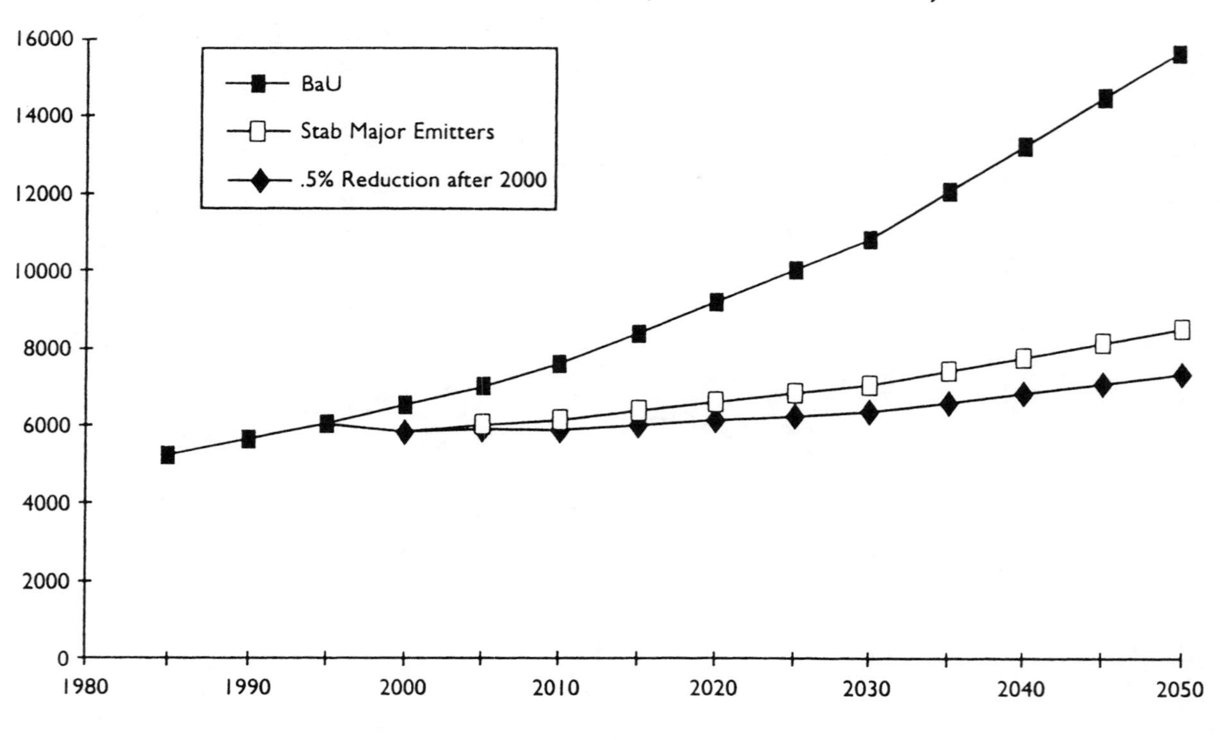

FIGURE 11.9 World carbon dioxide emissions (in mil. tons of Carbon).

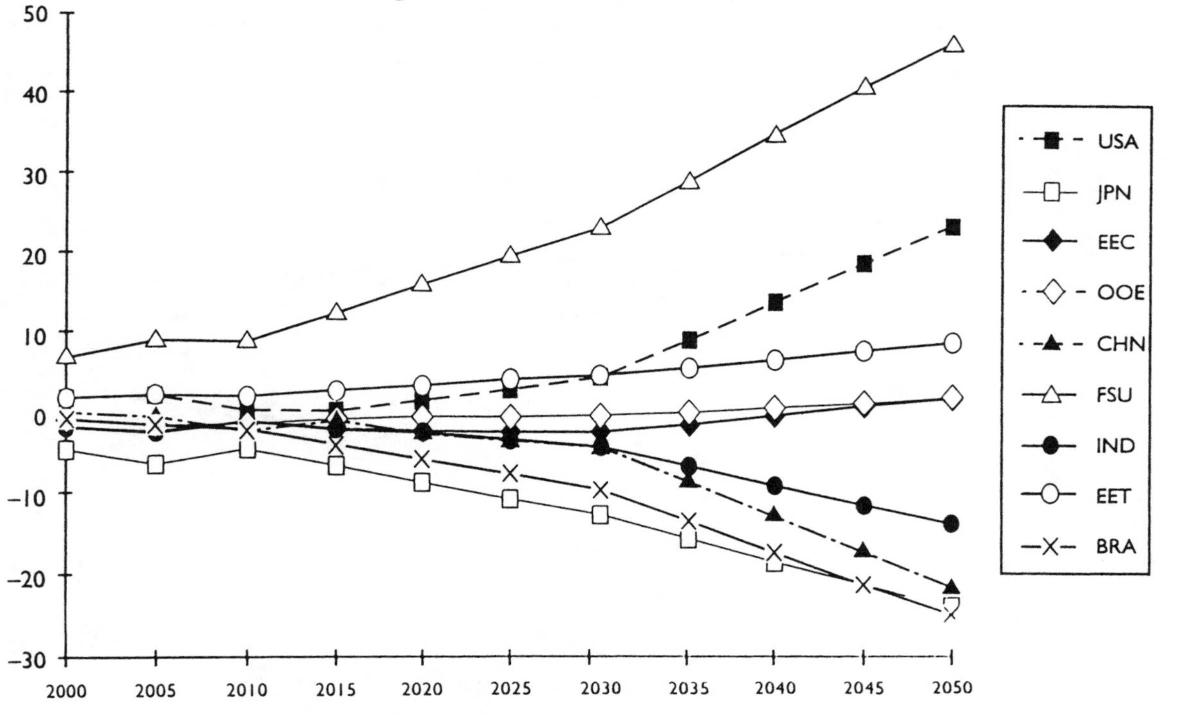

FIGURE 11.10 Emission rights trade: Grandfather allocation.

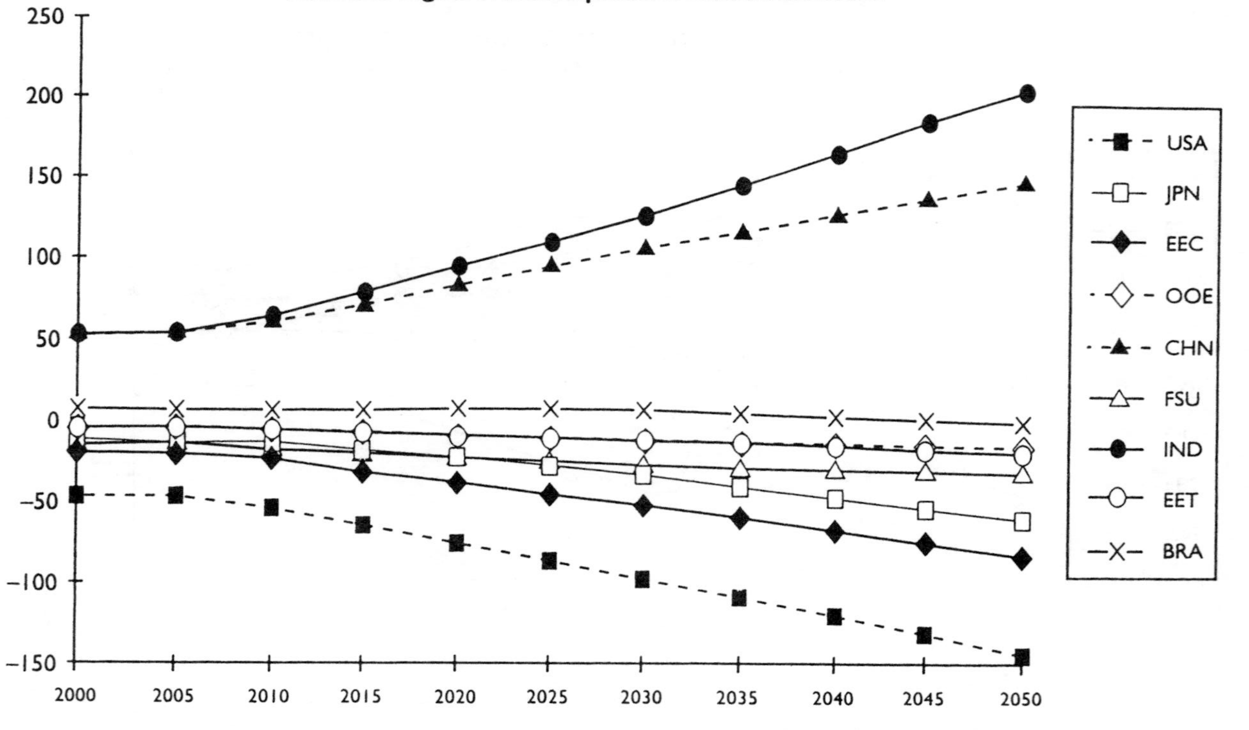

FIGURE 11.11 Emission rights trade: Population-based allocation.

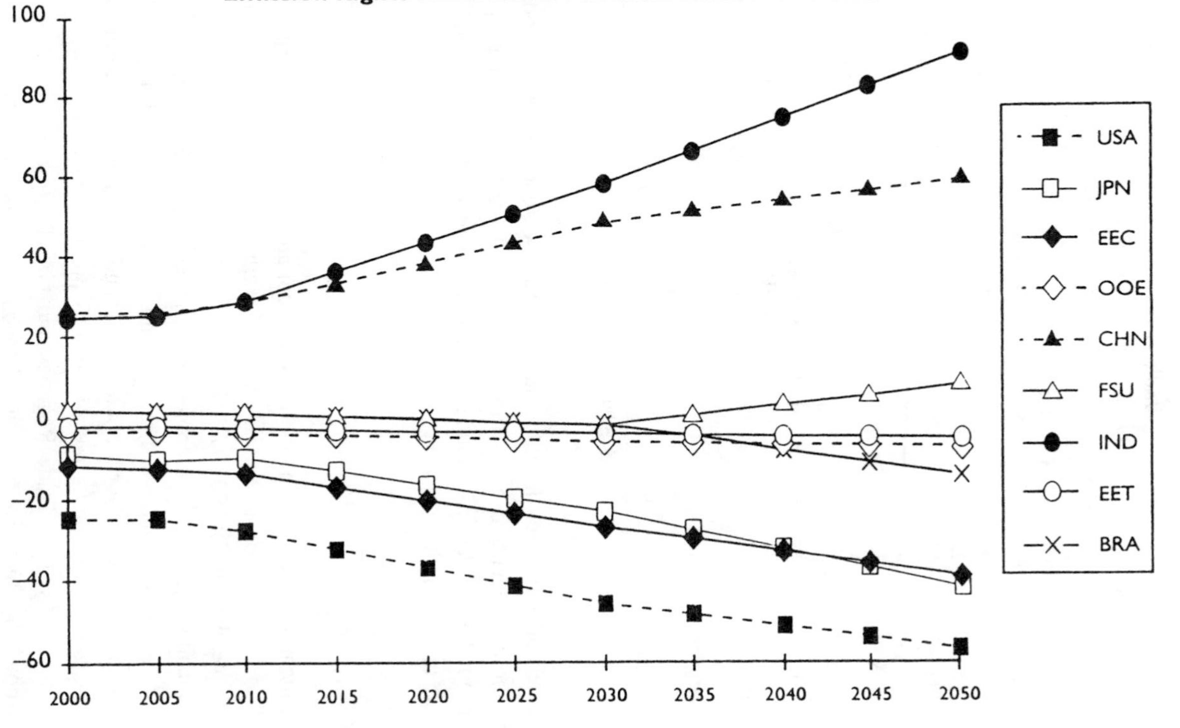

FIGURE 11.12 Emission rights trade: Mixed allocation.

Table 11.1
Real Income Loss over 2000–2050
(in percentage deviation relative to BaU)

	Indiv. Stab.	**Uniform Tax**	**Grandfathering**	**Pop. Based**	**Mixed**
USA	−0.79	−0.90	−0.76	−2.94	−1.84
JPN	−2.41	−1.24	−1.83	−2.84	−2.34
EEC	−1.23	−1.16	−1.22	−3.13	−2.19
OOE	−0.58	−0.55	−0.54	−1.53	−1.04
EEX	−3.39	−0.83	−0.78	0.09	−0.39
CHN	−3.88	−3.47	−4.14	6.02	1.04
FSU	−1.42	−2.66	1.08	−7.13	−2.92
IND	−2.61	−2.00	−2.94	14.62	7.00
EET	−0.33	−1.09	0.81	−5.94	−2.51
DAE	−0.29	0.16	0.20	−0.19	−0.05
BRA	−1.60	−1.78	−4.40	−0.55	−2.45
ROW	−0.40	−0.01	0.05	0.21	0.12
World	−1.65	−1.16	−1.17	−1.06	−1.07

Note: World Emission Stabilized at 1990 level after 2000.

the same dollar invested in industrial nations. If carbon abatement leads to decreased investment, then it is more efficient to decrease investment by one dollar in industrial nations, as the economic loss is relatively lower than if decreasing investment in industrial nations.

References

1. Chichilnisky, G. "Economic Development and Efficiency Criteria in the Satisfaction of Basic Needs." *Applied Mathematical Modelling* 1, no. 6 (September 1977): 290–97.
2. Chichilnisky, G. "Development Patterns and the International Order." *Journal of International Affairs* 31, no. 2 (fall/winter 1977): 275–304.
3. Chichilnisky, G. "North-South Trade and the Global Environment." *American Economic Review* 84, no. 4 (September 1994): 427–34.
4. Chichilnisky, G. "Trade Regimes and GATT: Resource Intensive vs. Knowledge Intensive Growth." *Economic Systems* 20 (1996): 147–81. Also published in *Handbook on the Globalization of the World Economy,* ed. Levy-Livermore (Cheltenham: Edward Elgar), chap. 10, pp. 226–49.
5. Chichilnisky, G. "The Knowledge Revolution." In *New Economy.* London: Institute for Public Policy Research, 1997, pp. 107–11.
6. Chichilnisky, G. "The Knowledge Revolution." *Journal of International Trade and Economic Development* 7, no. 1, (1998): 39–54.
7. Chichilnisky, G. "Property Rights and the Dynamics of North-South

Trade." *Structural Change and Economic Dynamics* 4, no. 2 (December 1993): 219–48.

8. Chichilnisky, G. "The Knowledge Revolution and Its Impact on Consumption and Resource Use." 1998 Human Development Report, United Nations Development Program, New York, 1998.
9. Chichilnisky, G., and G. Heal. "Global Environmental Risks." *Journal of Economic Perspectives,* Special Issue on the Environment (fall 1993): 65–86.
10. Chichilnisky, G., and G. Heal. "Who Should Abate Carbon Emissions? An International Perspective." *Economic Letters* (spring 1994): 443–49. (Also chapter 7 of this volume.)
11. Chichilnisky, G., and G. Heal. "Markets with Tradable CO_2 Emission Quotas: Principles and Practice." Working Paper No. 153, Economics Department, OECD, Paris, 1995.
12. Chichilnisky, G., and G. Heal. "Economic Returns from the Biosphere." *Nature* 391 (February 12, 1998): 629–30.
13. Chichilnisky, G., G. Heal, and D. Starrett. "International Markets with Emissions Rights of Greenhouse Gases: Equity and Efficiency." Publication No. 81, Center for Economic Policy Research, Stanford University, fall 1993; see also chapter 3 in this volume.
14. Chichilnisky, G. "Development and Global Finance: The Case for an International Bank for Environmental Settlements (IBES)," United Nations Educational, Scientific and Cultural Organization (UNESCO) and UN Development Program (UNDP), Discussion Paper No. 10, September 1996. See also *Sustainability and Global Environmental Policy: New Perspectives* (ed. A. Dragun and K. Jakobsson), Edward Elgar, Cheltenham, 1997, pp. 249–78.
15. *World Bank: World Development Report 1994.* Oxford: Oxford University Press, 1994.
16. Chichilnisky, G. "North-South Trade and the Global Environment" *American Economic Review* 84, no. 4 (September 1994): 851–74.
17. Chichilnisky, G. "Catastrophe Bundles Can Deal with Unknown Risks." *Bests' Review,* February 1996, p. 44–48.
18. Chichilnisky, G. "Technology Transfer and Emissions Trading: A Win-Win Approach to the Kyoto Protocol." Awarded a United Nations Foundation Grant to Columbia University's Program on Information and Resources, September 1998.
19. Chichilnisky, G. "A Comment on Implementing a Global Abatement Policy: The Role of Transfers." Paper presented at the June 1993 OECD Conference on the Economics of Climate Change, published in *The Economics of Climate Change,* ed. Tom Jones (Paris: OECD, 1994), pp. 159–70.

Chapter 12

The Clean Development Mechanism: Unwrapping the "Kyoto Surprise"[1]

Jacob Werksman

12.1 Introduction

Proposals that led to the adoption of the Clean Development Mechanism (CDM, Article 12) of the Kyoto Protocol[2] emerged late in the negotiating process, and consensus on the final text developed with unprecedented speed. The speed of this process, and the centrality of the CDM in brokering the final outcome of Kyoto, have led the chairman of the negotiations to refer to Article 12 as the "Kyoto Surprise."[3] Aspects of the CDM are undeniably innovative and have the potential to take the climate regime and indeed international law into uncharted territory. However, many of the CDM's core concepts can be traced directly to principles and mechanisms that have been discussed within the climate regime since the outset of the negotiations of the Framework Convention.[4]

In essence the CDM will facilitate a form of project-based joint implementation, governed by a multilaterally agreed set of rules and operating under the supervision of an intergovernment body. Annex I (industrialized) parties that

---

[1]Another version of this appeared in volume 7, issue 2, of the *Review of European and International Environmental Law.*

[2]The Kyoto Protocol to the United Nations Framework Convention on Climate Change (FCCC), adopted December 11, 1998. Uncorrected text at 37 ILM 22 (1998); the corrected text, and most other official documents cited in this chapter, can be found at the Web site of the secretariat to the FCCC at http://www.unfccc.de.

[3]Remarks by Ambassador Raúl Estrada y Oyuela, From Kyoto to Buenos Aires: Technology Transfer and Emissions Trading, a conference held at Columbia University, New York, 24 April 1998.

[4]UN Framework Convention on Climate Change, 31 ILM 849 (1992), entered into force March 21, 1994.

invest in projects in non–Annex I (developing) parties may use the greenhouse gas emission reductions accruing from such projects to offset a part of their commitments to limit or reduce their emissions under Article 3 of the protocol. Proponents of joint implementation see such investments as providing for win-win opportunities, whereby industrialized countries are allowed to achieve their commitments through the most cost-effective and flexible means and developing countries gain access to financial resources and clean energy technologies. However, as Article 12 took shape and gained momentum, various delegations sought to accommodate within the CDM the means for achieving a range of other objectives.

This chapter was prepared for the conference "From Kyoto to Buenos Aires: Technology Transfer and Emissions Trading," organized at Columbia University in April 1998 as a follow-up to the Conference of the Parties that led to the Kyoto Protocol. The conference, chaired by Raúl Estrada-Oyuela, explored the conceptual roots of different aspects of the CDM, including the pilot phase for activities implemented jointly, the functioning of the convention's financial mechanism, efforts to secure funding for adaptation, and the negotiations on the regime's compliance provisions. The negotiating history of Article 12 is reviewed here with reference to the specific textual proposals by both industrialized and developing countries that provided the elements of what would become the CDM. This is then followed by a close textual analysis of Article 12, which reveals significant ambiguities, and wide-ranging perceptions on how the CDM should evolve.[5]

12.2 Conceptual Roots

12.2.1 Project-Based Joint Implementation — Although the term *joint implementation* is not defined in the UN Framework Convention on Climate Change (FCCC),[6] it has been used to refer to two distinct but related concepts:

[5]The relationship with other proposals, such as the proposal for an International Bank of Environmental Settlements (IBES), is discussed in chapter 11. See "Development and Global Finance: The Case for an International Bank for Environmental Settlements (IBES)," Discussion Paper Series No. 10, United Nations Educational, Scientific and Cultural Organization (UNESCO) United Nations Development Programme, Office of Development Studies, September 1996. Also see "Development and Global Finance: The Case for an International Bank for Environmental Settlements," in *Sustainability and Global Environmental Policy: New Perspectives,* ed. A. K. Dragun and K. M. Jakobsson, Cheltenham: Edward Elgar, 1997, pp. 249–78.

[6]The two major references to the concept in the convention appear in Article 4.2(a), which anticipates that Annex I (developed) parties "may implement . . . policies and measures jointly with other Parties," and Article 4.2(d), which requires the Conference of the Parties, at its first session, to take decisions regarding "criteria for joint implementation." The text of the convention and all official documents cited in this article can be found on the secretariat's Web site at http://www.unfccc.de.

(1) project-based joint implementation, which would allow Annex I countries to obtain carbon offsets or credits toward their emissions reduction targets in exchange for investment in mitigation projects abroad in either Annex I or non–Annex I parties where the costs of such investments are lower, and (2) a system of tradable emissions allowances that, once allocated between parties or groups of parties, can be traded subject to a set of prescribed rules. Both forms of joint implementation were conceived to enable Annex I parties to achieve their commitments to reduce greenhouse gas emissions in a more cost-effective manner and to encourage transfers of financial resources and/or technology between parties. Both forms, however, have provoked concern from parties and observers, who argue that joint implementation shifts the responsibility, if not the cost, of undertaking emissions cuts from developed to developing countries and that this shift in responsibility could make it more difficult to ensure compliance with emissions reduction obligations.

Proponents of joint implementation have argued that such arrangements are legally possible with no justification additional to the text of the convention. An early launch of joint implementation was, however, constrained by the absence of mutually agreed criteria for joint implementation, which, the convention provides, were to be agreed by the First Conference of the Parties (COP1). Nonetheless, soon after the convention entered into force, potential investor countries, most notably the United States and Norway, began experimenting with projects in developing and transition countries designed to demonstrate the feasibility of generating carbon offsets. However, in the context of uncertainty about whether and under what criteria such offsets would be credited by the COP and in the absence of clearly quantified legally binding commitments, there was little incentive to do more than experiment.

Pilot Phase for Activities Implemented Jointly

As a result of these legal and political uncertainties, little had been done in time for COP1 to develop either the methodologies or the confidence among the critics of joint implementation that would be necessary to build a consensus decision on criteria for joint implementation. Instead, after intense negotiations, COP1 established a pilot phase for activities implemented jointly (AIJ).[7] The purpose of the pilot phase was to provide a more transparent and coherent basis for testing the feasibility of JI.

---

[7]Report of the Conference of the Parties on its First Session, FCCC/CP/1995/7/Add.1, April 1995, Decision 5/CP.1.

Constructive ambiguities built in to the pilot phase decision, including the newly coined acronym AIJ, allowed JI proponents to claim that the concept of project-based carbon offset investments had been approved in principle while skeptics could maintain that joint implementation was still on trial. The core of the AIJ decision clearly tipped the balance toward the skeptics by denying AIJ investors the possibility of obtaining credit, even retroactively, for any emissions reductions achieved through investments made during the AIJ pilot phase.

The negotiations of the AIJ decision and the operation of the pilot phase did, nonetheless, help to flush out and to elaborate a number of issues of principle and of practicality that influenced the development of Article 12 of the protocol and that will be critical to the ongoing discussions on the CDM. Perhaps most crucially, the COP1 negotiations resolved that, despite the references to JI in the convention, decisions on whether and on what basis credit for investments could be offset against commitments could not be taken unilaterally or through bilateral agreement between an investor and a host party. Such decisions could only be taken by the COP and would thus necessarily depend on the overwhelming support of the numerically dominant developing country parties.

Despite the unavailability of credit during the AIJ pilot phase, JI proponents made significant investments in demonstration projects, which at COP3 totaled 122. These were carried out among a very limited number and range of parties,[8] primarily through bilateral initiatives, such as the United States Initiative on Joint Implementation (USIJI) program, and the Norwegian/World Bank AIJ program. At the time of COP3, only the United States, Norway, and the Netherlands had developed AIJ projects with partners outside Annex I.[9]

The COP and its Subsidiary Body on Scientific and Technological Advice (SBSTA) developed a uniform reporting format for AIJ. The review of these reports by the FCCC secretariat and the SBSTA allowed a number of significant political and methodological issues to emerge that will help inform discussions on CDM development. Many supporters of AIJ, both North and South, have recognized rigorous reporting as essential to the successful use of JI as a means of achieving real net reductions in global greenhouse gas emission.

---

[8]The secretariat's most recent analysis of AIJ reports indicates a significant increase in the number and geographical spread of projects. With regard to geographical distribution, 29 of the 122 projects are based in Latin America (nine of which are hosted by Costa Rica), five are in Africa, and nine are in Asia. See FCCC/SB/1995/5.

[9]FCCC/SBSTA/1997/INF.3.

Thus far, resistance to rigorous reporting standards for AIJ projects has come from a number of developing countries that are concerned that mechanisms for monitoring compliance of individual AIJ projects are a first step toward extending significant emissions reduction and reporting requirements to developing countries as a group. The Group of 77 developing countries (G-77), which, in the climate process, provides the primary negotiating forum for non–Annex I countries, has historically resisted detailed reporting on greenhouse gas emissions as being too intrusive an imposition on national sovereignty. Although none has stated so openly, some developed countries might also resist rigorous reporting on AIJ, as it will necessarily increase the transaction costs involved in each project and might reveal fundamental impracticalities in the approach that render it less attractive.

Under the evolving drafts of the AIJ uniform reporting format,[10] AIJ partners must demonstrate the following for each project:

1. Environmental additionality, that is, that the AIJ project brings about real, measurable, and long-term environmental benefits related to the mitigation of climate change that would not have occurred in the absence of the project, and
2. Financial additionality, that is, that the resources from the Annex I investor are additional to the financial obligations of the Annex I party under the convention as well as to current official development assistance flows.

Demonstrating that the AIJ investment has yielded net additional environmental benefits thus requires the AIJ partners to construct a counterfactual baseline or reference case that describes what the host country would have done in the absence of the AIJ project. Furthermore, project proponents wanted to discourage the problem of leakage, whereby emissions increases within the host country but outside the scope of the project might wipe out the project's environmental benefits. Preventing or accounting for leakage might require a baseline to be constructed that would assess potential emissions on a countrywide basis. Such counterfactual determinations are inherently difficult and, especially when left to bilateral negotiation, take place in a context in which both the

[10] An initial draft of the AIJ Uniform Reporting Framework was presented to the parties in FCCC/SBSTA/1997/3. A modification of this format, contained in FCCC/SBSTA/1997/4, was adopted by COP3, in Decision 10/CP/3. A draft version of the URF is proposed in FCCC/SB/1999/5/Add. 1, p. 13.

investor and the host share strong incentives to overstate the baseline emissions scenario in order to inflate the offset credited to the project.[11]

With regard to financial additionality, pilot phase AIJ and the uniform reporting framework were designed to ensure that developed countries did not use AIJ investments in place of the investment in developing country capacity that they are already required to make under the convention's financial mechanism.[12]

Just prior to Kyoto, the FCCC secretariat undertook an analysis of the AIJ reports received and confirmed that parties were struggling with these methodological challenges and producing inconsistent results.[13]

End of the Pilot Phase, Start of the Protocol

The AIJ negotiations revealed the depth of skepticism with which many developing countries view JI. Their resistance to JI, in the face of political pressure and the offer of financial incentives, might best be summarized as a combination of concerns that fully operational JI for credit would be used to constrain their development choices. Unequal bargaining positions in bilateral JI negotiations could allow Annex I investors to impose new conditionalities for access to financial resources and technology transfer, to promote the projects that were not in the national interest and that could divert resources from official development assistance (ODA) and GEF resources.[14,15]

---

[11] For a discussion of the methodological challenges associated with AIJ and the protocol's flexibility mechanisms, see Activities Implemented Jointly: Partnerships for Climate and Development (IEA/OECD:1997); J. Heister, "Baselines and Indirect Effects in Carbon Offsets Projects: A Guide for Decision-Making," Draft 20, World Bank, January 1998.

[12] When seeking to determine whether contributions to the GEF were, as the convention requires, "new and additional," an independent panel of experts concluded that until international rules were developed, such a determination was not possible. G. Porter, R. Clémençon, W. Ofosu-Amaah, and M. Philips, "Study of GEF's Overall Performance" (GEF 1998) (hereafter GEF Study). During the pilot phase, AIJ projects are to be funded with resources additional to those provided by Annex II parties in fulfilment of their financial obligations within the framework of the financial mechanism and in addition to current official development assistance (ODA) flows. These flows are, however, notoriously difficult to monitor and compare, and it is not clear how developed country parties will be able to establish, in the context of declining overall flows of ODA, that investments in AIJ are additional to resources that would have or should have been committed to the GEF or to other sources of ODA.

[13] FCCC/SBSTA/1997/INF.3.

[14] FCCC/SBSTA/1997/MISC.5.

[15] The AIJ pilot phase continues, and, at least until the entry into force of the protocol, its fate will remain linked to the obligations and the institutions of the convention rather than the protocol. Efforts will no doubt be made to fold AIJ projects involving developing countries into the CDM. However, until these issues are resolved, in the interim period before the protocol and the CDM begin to operate, the practical experience gained through AIJ will continue to influence the development of methologies and procedures for the CDM.

12.3 The Global Environment Facility

The Global Environment Facility (GEF) has served, since the adoption of the convention, as the operating entity responsible for matching eligible projects in developing country parties with funds provided by Annex II parties under the convention's financial obligations. The GEF will be of interest to those working on the CDM as both a forerunner and a potential competitor for CDM projects. The methodologies developed by the GEF over the past five years of its operation to calculate the global environmental benefits generated by its investments might provide a basis for measuring the value of carbon offsets accruing from a CDM investment.

The GEF, which also serves as the financial mechanism for the other major Rio treaty, the Convention on Biological Diversity, represents what can be termed the UNCED approach to financing treaty implementation in developing countries.[16] Following the principle of common but differentiated responsibility, the Annex II parties (the wealthier Annex I countries) are required to provide new and additional funds to cover the agreed full incremental costs of measures undertaken by developing country parties to implement the convention. The extent to which developing countries are expected to fulfil their commitments is explicitly conditioned on the compliance of developed countries with their financial obligations.[17]

To limit the scope of their financial commitment and to help ensure the most effective use of the GEF's resources, Annex II parties encouraged the development of methodologies for calculating the incremental cost of greenhouse gas mitigation projects. In theory, under an incremental cost discipline, the GEF funds only that element of a project that results directly in the reduction of greenhouse gas emission, yielding thereby a global environmental benefit. Under this methodology a project proponent must describe a baseline scenario of the activity that would have taken place in the host developing country but for the GEF investment. The GEF then provides the funding that makes the alternative or additional climate-friendly activity possible.

Thus, both GEF projects and project-based carbon offset activities developed under the CDM will require the design and identification of projects or project activities that can be demonstrated to result in identifiable emissions reductions.

---

[16]In fact GEF concepts trace directly to financial arrangements under the 1997 Montreal Protocol on Substances that Deplete the Ozone Layer, 26 ILM 1550, and its Multilateral Fund. See J. Werksman, "Consolidating Governance of the Global Commons: Insights from the Global Environment Facility," *Yearbook of International Environmental Law* 6, no. 27 (Oxford, 1995).

[17]UNFCCC, Articles 4.3, 4.7.

A recent assessment of the GEF's overall performance, commissioned from an independent review team, highlighted the challenges that the GEF continues to face in applying the incremental cost methodology. Although the GEF's approach has improved and become more flexible over time, the review team noted that the "present process of determining incremental costs has excluded the participation of recipient country officials in most cases, because of the lack of understanding of the concept and methodologies." [18] If project-based JI is to attract the support of host countries, the CDM will have to overcome similar challenges to produce a methodology that is transparent and practicable. The experience thus far with the GEF project cycle indicates that the process of identifying and designing projects that can be claimed with any confidence to generate emissions reductions that would not otherwise have occurred can be fraught with political and methodological difficulties.

Developing countries, the primary recipients of GEF funds have, since Rio, consistently expressed their disappointment in the GEF, reflected most clearly in their refusal to confirm the GEF as the permanent operating entity of the convention's financial mechanism. This disappointment stems from the perceived inadequacy of GEF funding levels, the slowness of the GEF project cycle, and the continued dominant influence of donors and the World Bank in shaping GEF policy. Although the GEF has been effectively confirmed by the protocol and the GEF Council, to play the same role in funding the protocol that it has played in funding the convention, its rocky beginnings opened an opportunity for an alternative funding mechanism and helped make the CDM possible.[19]

12.3.1 Funding Adaptation — Article 4.4 of the convention requires Annex II parties to assist those developing country parties most vulnerable to the adverse effects of climate change to meet the costs of adapting to those adverse effects. Annex II parties have, however, been concerned about the potentially unlimited cost associated with this obligation and the implication that compensating countries for the impacts of climate change concedes liability for their historical role in raising atmospheric greenhouse gas concentrations. Consequently, Annex II stiffly resisted links between Article 4.4 and the convention's financial mechanism. The GEF's focus on incremental cost financing was interpreted by donors to preclude it from funding activities other than those that generate global environmental benefits. Investments in coastal zone manage-

[18] GEF Study.

[19] Kyoto Protocol, Article 11(2)(b); "The New Delhi Statement of the First GEF Assembly," April 3, 1998, available at http://www.gefweb.com.

ment, strengthening sea defenses, or preparing for shifts in agricultural patterns have been viewed as generating domestic benefits outside the GEF's ambit.

At COP1 delegations from developing countries especially vulnerable to the adverse effects of climate change overcame the resistance of major donor countries and secured the endorsement of policies, eligibility criteria, and program priorities that ensure that funding will be provided for a first, limited category of adaptation projects (Stage I projects).[20] Since then the GEF Council has adopted an operational strategy that provides more detailed criteria for the funding of Stage I projects[21] and has approved a handful of projects.

In Stage I, developing country parties especially vulnerable to the impacts of climate change are eligible for full cost financing of adaptation activities related to preparing their national communications and national climate change programs, as required under Articles 4.1 and 12 of the convention.[22,23]

The absence of any meaningful source for adaptation funding under the convention opened a further opportunity for building support for an alternative funding mechanism. Emerging proposals from the CDM had the potential to generate income that could be earmarked for adaptation, that would be free from the GEF's incremental cost analysis, and that would not necessarily entail additional financial resources from governments.

12.3.2 Compliance — The history of the treatment of compliance issues under the climate change regime is reflected in the text of Article 13 of the convention and the subsequent and ongoing work of the Ad Hoc Group on Article 13 (AG-13). The majority of delegations did not support the inclusion of a robust mechanism for enforcing compliance with the convention's soft

[20]Decision 11/CP.1, Initial Guidance on Policies, Programme Priorities and Eligibility Criteria to the Operating Entity or Entities of the Financial Mechanism.

[21]*Operational Strategy* (Washington, D.C.: GEF), February 1996, pp. 38–39.

[22]These enabling activities are limited in nature but can include funds for training, vulnerability assessment, and planning related to adaptation. The GEF's operational guidelines for the funding of enabling activities indicate a typical cost range of up to $350,000 per country for the entirety of the enabling activities. These funds would be expected to include not only Stage I adaptation costs but also costs of preparation and initial national communication.

[23]"Operational Guidelines for Expedited Financial Support for Initial Communications from Non-Annex I Parties to the United Nations Framework Convention on Climate Change," GEF/C.7/Inf,10/Rev.1, October 3, 1997. In approving this approach to expediting national communications, the GEF Council noted that "the financing amounts for the preparation of enabling activities have been developed on the basis of an average estimate used for planning purposes. However, the actual level of support will vary from country to country and with the content of the enabling activities." Joint Summary of the Chairs, GEF Council Meeting, April 2–4, 1996, "Appendix: Council Decisions, Decision on Agenda Item 5(b)."

and ill-defined obligations.[24] Negotiations since the convention entered into force have instead focused on the consideration of the establishment of a non-confrontational and facilitative multilateral consultative process for the resolution of questions regarding implementation of the convention.

However, the course of the protocol negotiations revealed that strengthened commitments and more sophisticated means for implementing those commitments would require a correspondingly more elaborate system for identifying noncompliance and for providing a range of incentives and disincentives for encouraging compliance.

The possibility that noncompliance by Annex I parties could, through the imposition of financial penalties, provide a source of revenue for development assistance proved very attractive to non–Annex I delegations. Establishing pre-set penalties or financial safety valves as remedies for noncompliance with or breach of an international treaty raises complex issues with regard to the nature of international legal obligations. Traditional concepts of state responsibility envision that international practice demands reparation for a breach that "as far as possible, wipe[s] out all the consequences of the illegal act and re-establish[es] the situation which would, in all probability, have existed if that act had not been committed."[25] Such consequences are difficult to prejudge.

Nevertheless, noncompliance, financial penalties, and a link to development assistance became the conceptual filter through which JI was perceived as acceptable to the majority of G-77 countries.

12.4 Negotiating History of Article 12

12.4.1 Initial Positions on Project-Based JI — Project-based JI between Annex I and non–Annex I parties was introduced from the outset of the protocol negotiations and was incorporated in the Negotiating Text by the Chairman (NTC).[26] These proposals ranged from absolute prohibitions on JI

[24]On the evolution of compliance mechanisms under the UNFCCC, see H. Ott, "Elements of a Supervisory Procedure under the Climate Regime," *Heidelberg Journal of International Law,* 56, no. 3 (1996), and J. Butler, "Establishment of a Dispute Resolution/Non-Compliance Mechanism in the Climate Change Convention," unpublished manuscript (on file with the author).

[25]Cherzow Factory (Indemnity) case, PCIJ, Ser. A, no. 17, p. 47, as cited in I. Brownlie, *Principles of Public International Law,* 4th ed. (Oxford: Clarendon Press, 1990).

[26]The Negotiating Text by the Chairman (FCCC/AGBM/1997/3/Add.1 and Corr.1), dated April 21, 1997, prepared by the chairman, with assistance from the secretariat, is a comprehensive document reflecting all submissions made by parties to date and structured in the form of a protocol and without attribution to the parties. Prepared both to assist the negotiations and to meet the convention's procedural deadline (Articles 15.2 and 17.2) requiring that any proposals for protocols or amendments to the conven-

(Iran),[27] to proposals that would have limited JI to Annex I parties only (EU),[28] to more detailed elaboration on the conditions under which non–Annex I countries would be entitled to participate in project-based JI (United States).[29]

Although the G-77/China position emphasized that "[e]ach party included in Annex I to the convention shall meet its QELROs through domestic action,"[30] individual members of the group began to rebel against an outright prohibition on JI. Most notable of these was the proposal of Costa Rica, a country with an active AIJ program and that would later play a key role in designing the CDM.[31]

The Consolidated Negotiating Text by the Chairman (CNT) was prepared prior to the last scheduled session of the Ad Hoc Group on the Berlin Mandate (AGBM) and reflected the chairman's assumptions as to the "thrust of deliberations in the Group to date." It supported the prevailing position of the European Union and of the G-77 and would have allowed project-based JI between Annex I parties only.[32]

12.4.2 The Brazilian Clean Development Fund — The basis for a breakthrough in the negotiations of project-based JI between Annex I and non–Annex I parties arrived with the submission by the government of Brazil of the Proposed Elements of a Protocol.[33] This sweeping proposal sought to radically redefine the climate regime from the ground up.[34] Drawing inspiration from

---

tion be submitted six months prior to the COP at which they are proposed for adoption. Of particular importance to the negotiations on the CDM in that with it the negotiators recognized that "whilst proposals additional to this negotiating text may be put forward, these should be clearly derived from the submissions already within it and should not introduce substantially new ideas."

[27] CNT, para. 139.

[28] Ibid., para. 140.

[29] Ibid., para. 143.

[30] Ibid., para. 121.4.

[31] Ibid., paras. 147–147.6.

[32] The Consolidated Negotiating Text by the chairman (FCCC/AGBM/1997/7), dated October 13, 1997, prepared just prior to the commencement of AGBM-8, was the first effort to produce a text that had the appearance of a protocol. Although significantly bracketed, and prefaced with the caveat that it was offered "without prejudice to" the NTC and the original proposals from parties contained in the relevant MISC DOCs, the chairman's assumptions as to the "thrust of deliberations in the Group to date" were employed to substantially narrow the options previously reflected in earlier compilations.

[33] FCCC/AGBM/1997/MISC.1/Add.3, p. 3.

[34] The inspiration for this proposal seems to have come from a number of sources. Its strong foundation in climate science and IPCC modelling ties it directly to Brazil's chief negotiator in the AGBM process, Dr. Luiz Gylvan Meira Filho, president of the Brazilian Space Agency and IPCC lead author. Dr. Meira Filho was given the responsibility for chairing the informal Contact Group on what became the CDM and is widely credited for successfully steering it through the negotiations. The economic aspects of the proposal, and especially the aspects that allow trading between Annex I parties, bear some resem-

IPCC climate models and emissions scenarios, the Brazilian protocol sought to introduce a science-based objectivity into the negotiations. The protocol's overall objective was to define a future level of effective emissions that could be tolerated from Annex I countries on the basis of the predicted impact of these emissions on global mean surface temperatures. An effective emissions ceiling for the combined emissions of Annex I countries for each of four five-year budget periods, running from 2001 to 2020, was proposed. Differentiated individual effective emissions ceilings were then to be allocated amongst Annex I parties on the basis of the relative fraction of effective emissions attributable to each Annex I party from modeled emissions projections.

For the purposes of the development of the CDM, the most important element of the Brazilian proposal was the introduction of a compulsory contribution, or a financial penalty for noncompliance, to be assessed against each Annex I party that had exceeded its effective emissions ceiling at the end of its budget period. The penalty was to be contributed to a non–Annex I clean development fund for use in funding climate change projects in developing countries. The size of the penalty was designed to correlate to $10 for every tonne of carbon equivalent that the Annex I party had exceeded its ceiling. This amount was estimated to reflect the likely cost of achieving an equivalent level of emissions reductions through the "implementation of non-regrets [*sic*] measures by non–Annex I parties."[35]

Brazil went further, proposing an objective basis for distributing the funds among non–Annex I parties. First, funding would be provided to non–Annex I parties in response to a voluntary application subject to the appropriate regulations approved by the COP. Second, the eligibility of each non–Annex I party for funding would be capped at a level on the basis of its relative responsibility for effective emissions during the preceding budget period. An appendix divided potential proceeds from a clean development fund into shares based on projected emissions from 1990 to 2010, ranging from China at 32% to Niue at .00005%.[36] Finally, up to 10% of the Brazilian clean development fund would be made available to non–Annex I parties for use in adaptation projects.

Critics of the Brazilian protocol doubted that such a radical restructuring of the regime could be managed in the months left before Kyoto and noted that

blance to proposals for a "Green Bank" being made by Professor Graciela Chichilnisky of the Columbia University Program on Information and Resources. See Chichilnisky, "Development and Global Finance: The Case for an International Bank for Environmental Settlements," 10 UNDP Discussion Paper Series, UNDP, 1997.

[35]FCCC/AGBM/1997/MISC.1/Add.3, p. 24.

[36]Ibid., p. 54.

the logic of objective effectiveness resulted in a regime that penalized the large emitters in Annex I through higher commitments while rewarding the largest non–Annex I emitters with access to the largest share of the funds. There was, however, enough in the proposal to prove selectively attractive to a wide range of parties.

The first significant advance came with the formal endorsement by the G-77 and China of a central aspect of the Brazilian proposal. In a submission to the final session of the AGBM, the G-77 endorsed the establishment of a clean development fund as a means of enforcing compliance with Annex I commitments while generating revenues for development assistance. The Brazilian proposal was stripped to its essentials and incorporated into a position of the G-77 and China as follows:

> A Clean Development Fund shall be established by the Conference of the Parties to assist the developing country Parties to achieve sustainable development and contribute to the ultimate objective of the Convention. The Clean Development Fund will receive contributions from those Annex I Parties found to be in noncompliance with its QELROs under the Protocol.[37]

The United States embraced the flexibility the Brazilian proposal appeared to offer to Annex I countries having difficulty meeting their commitments at home. Characterizing the proposal as a trading system and a flexible financing instrument, the head of the U.S. delegation expressed the view that the proposal for a clean development fund and its endorsement by the G-77 represented a significant basis for hope in the approach to Kyoto.[38]

This broad-based support for some variation on a clean development fund (although subject to a very wide range of interpretation) was enough to convince the AGBM chairman to include the G-77 paragraph in the Revised Text Under Negotiation (RTUN) that went forward to Kyoto.[39,40]

---

[37] Ibid., Add.6, p. 16.

[38] Reuters News Service, "Delegates Say Prospects Brighten for CO_2 Treaty," November 10, 1997.

[39] Revised Text Under Negotiation (RTUN) FCCC/CP/1997/2, although the G-77 formulation of the clean development fund was received on October 22, 1997, well after the convention's June 1 deadline for substantially new submissions.

[40] RTUN, p. 9, n. 4; p. 18, n. 13. Significantly, however, the RTUN continued to reflect a resistance to project-based joint implementation between Annex I and non–Annex I parties. There was no provision for the calculation or transfer of emissions reduction credits that might result from such a fund. Instead, the G-77 text and its placement in the RTUN maintained its emphasis as a means of enforcing compliance.

Thus, just prior to the third session of COP3 the context was set for an exploration of how views of such diverging emphasis could somehow coalesce in the creation of a mechanism that would perform such a variety of functions.

12.4.3 The Kyoto Crucible — Work on what would become the clean development mechanism began almost immediately as delegations arrived in Kyoto. An informal contact group, under the chairmanship of Brazil, was established by the Committee of the Whole (COW) in the first hours of the negotiations to discuss the clean development fund and other financial issues.[41] The brief history of the negotiations in Kyoto can be characterized as a struggle that merged the U.S.-backed proposals for project-based JI and G-77 proposals for a fund fed by compliance penalties.

The initial response from the European Union was to view the emerging CDM with suspicion. As promoted by the United States, the CDM ran counter to the European Union's position against project-based JI with parties without commitments. The version supported by the G-77 was perceived by the European Union as creating a new institution that would threaten the continued viability of the GEF as the main source of convention funding.[42]

G-77 emphasis on the compliance aspects of the clean development fund became difficult to maintain once the negotiations began to divide into smaller contact groups. Compliance, and any role a clean development fund might play in it, was assigned to a sub-group on institutional aspects of the Protocol. This sub-group was dominated by Annex I Parties, discussing potential consequences for themselves of failing to meet their commitments. Text was introduced that would have channelled financial penalties into a clean development fund.[43] However, it became apparent that it would not be possible to agree what specific binding consequences might flow from a determination of non-compliance, and the direct link between compliance and the fund dissolved.[44]

This sidetracking of compliance led the contact group on a clean development fund to focus on the role such a mechanism might play in facilitating project-based JI. In the course of two days of negotiation, the original G-77 proposal evolved from a single paragraph bolted on to the Article on Annex I

---

[41] *Earth Negotiations Bulletin* 12, no. 68 (December 2, 1997); author's notes.

[42] *Earth Negotiations Bulletin* 12, no. 71 (December 5, 1997).

[43] See FCCC/CP/1997/CRP.2, 7 December 1997, Article 18. Alternative A. The CRP (conference room paper) series of documents were issued in Kyoto by the chairman, to reflect and consolidate progress from the various working and contact groups during the negotiations.

[44] See FCCC/CP/1997/CRP.4, 9 December 1997, Article 19.

commitments[45] to a freestanding Article of ten paragraphs, substantially in the form it would be adopted in the protocol.[46]

Within 48 hours, the following basic principles and design features for the CDM were agreed:

1. The informal contact group defined a mechanism rather than establishing a fund, reflecting the CDM's primary role as a processor of transactions rather than as a depository of financial resources and assuaging, in part, concerns about the proliferation of international institutions and threats to the role of the GEF as the regime's financial mechanism.
2. It was agreed that credit for reductions resulting from CDM investments made from 2000 onward can be offset against the part of commitments of investor countries, resolving the main point of principle that had been left hanging by the AIJ pilot phase.
3. New institutional features emerged, including an executive board and a role for the meeting of the Conference of the Parties serving as the meeting of the parties (COP/MOP). This provided multilateral, intergovernment supervision in response to G-77 concerns about the lack of fairness and transparency that many felt had characterized bilateral AIJ transactions.
4. General criteria were agreed to provide a basis for certifying emissions reductions resulting from CDM projects, which reflected many of the same principles with regard to the need for country-driven projects and for environmental additionality that had been agreed in the AIJ pilot phase.
5. The task of adopting more specific procedures for auditing and verifying emissions reductions was assigned to the COP/MOP, reflecting ongoing concerns from a wide range of delegations that CDM transactions might be open to abuse.
6. A role in the operation of the CDM for operational entities and private and/or public entities outside the convention/protocol institutions was agreed, in principle. This left open the possibility for the direct involvement of international institutions and the private sector.
7. The operation of the CDM would be expected to generate funds to cover administrative expenses, thus helping to assuage concerns about the proliferation and the costs of new international institutions.

---

[45] See FCCC/CP/1997/CRP.2, Article 3(19).

[46] See FCCC/CP/1997/CRP.4, Article 14.

Table 12.1
Articles and Provisions

Article	Provision
12.1	Definition
12.2	Objective
12.3	The transaction
12.4	Governance
12.5	Principles for the certification of emissions reductions
12.6	Project finance
12.7	Auditing and verification
12.8	Administrative expenses and adaptation costs
12.9	Involvement of private and/or public entities
12.10	Banking of certified emissions reductions

8. A share of the proceeds from the operation of the CDM will be used to assist especially vulnerable developing countries meet the costs of adaptation.

12.5 Article 12 and the CDM

Under Article 12 (see Table 12.1), the CDM will facilitate a form of project-based JI between Annex I and non–Annex I parties, governed by a multilaterally agreed set of rules, and operating under the supervision of the Conference of the Parties serving as the meeting of the Parties to the Protocol (COP/MOP) and an executive board. Emissions reductions accruing from project activities carried out in non–Annex I parties, once certified under agreed principles, may be used by Annex I parties to contribute to compliance with their emissions reductions obligations under Article 3 of the protocol.[47] Thus, agreement on Article 12 resolved a number of critical aspects of how the CDM will manage project-based JI between Annex I and non–Annex I parties to the protocol. However, many gaps remain to be filled, and the negotiating dynamic for the next stage of the development of the CDM remains fundamentally unchanged. This dynamic can now be characterized as pitting a market-based approach against an interventionist approach based on traditional public sector develop-

[47] Kyoto Protocol, Article 3.12.

ment assistance. Both approaches emphasize the need for a system capable of generating credible certified emissions reductions (CERs) but differ on the best means of achieving this.

A market-based approach relies on healthy competition in a transparent marketplace to provide the most efficient and effective means of encouraging hosts and investors to design credible CDM project activities. Once the intergovernmental process has set the rules on the types of project activities that will be eligible for certification, the private sector, which holds the capital and technology necessary to the CDM's success, would be entrusted with designing projects and would be entitled to hold and transfer CERs.

Interventionists are more skeptical of the private sector's ability to fulfil the CDM's stated purpose of assisting non–Annex I parties to achieve sustainable development. Such an approach emphasizes the need for the active involvement of public sector institutions, including home and host governments and international development institutions, in promoting the design of projects driven by broad-based policy concerns rather than market disciplines.

The debate is further complicated by the tension between those that want to see the CDM up and running quickly and with as low transaction costs as possible and those that remain cautious and are willing to increase costs in exchange for greater accountability. Parties at both ends of this spectrum place the CDM at risk either by undermining its credibility or by weighing it down with an overburdensome bureaucracy.

12.5.1 CDM Governance — Decisions on the operation of the CDM will ultimately be taken by its governing bodies (see Table 12.2). Article 12 entrusts the COP/MOP and an executive board with the general functions of guiding and supervising the CDM's operation. The division of labor between the two bodies is not entirely clear, and some refinements are likely to prove desirable. For example, the COP/MOP might want to delegate some of the more detailed work, such as the designation of operational entities, to the smaller, more focused body.

The Kyoto Protocol left undecided issues relating to the size, composition and modus operandi of the executive board (EB). The functions set out previously suggest that the EB will require a mixture of technical skills and political authority. The appropriate balance between these will depend, once again, on how interventionist the CDM is in the design, funding, and approval of project activities. The more actively involved it is in a project cycle, the greater its need for technical expertise.

The political composition of the EB will require consideration of balance between regions and/or between investor and host countries. Annex I countries

Table 12.2
CDM Governance

General Function	Specific Function	CDM
Governance	Provision of authority and guidance	COP/MOP
	Determination of part of commitment available for offset	COP/MOP
	Supervision	Executive board
	Elaboration of modalities and procedures for auditing and verification of project activities	COP/MOP
	Designation of operational entities to certify emissions reductions	COP/MOP
	Provision of guidance on the participation of private and/or public entities	Executive board
	Ensuring assessment of administrative and adaptation costs	COP/MOP
CER management	Certification of emissions reductions	Operational entity
	Auditing and verifying project activities	Undetermined (private/public entity?)
Project finance	Arranging funding of certified project activities	CDM (unspecified)

will no doubt argue against regional balance, as this inevitably leaves them with fewer seats than developing countries.[48] It must be kept in mind, however, that the larger the role played by the private sector in funding CDM projects, the weaker Annex I parties' claims for a disproportionate presence on the EB. If they are no longer in the position of donors, they have not bought their entitlement to a larger share of the vote.

Article 12 does not rule out the possibility that the function of the EB could be carried out by an existing institution that shares whatever design principles the parties agree to. Indeed Article 12(1) defines rather than establishes the CDM. This language is borrowed from Article 11 (Financial Mechanism) of the convention, where it was used to avoid the creation of a new institution, by allowing the GEF to operate the convention's financial mechanism. There would likely be considerable resistance from developing countries, which are underrepresented on the GEF Council, to authorizing the GEF to supervise the CDM.

[48]In UN practice, regional balance requires membership in multiples of five, representing Asia, Africa, Latin America, and the Caribbean (non–Annex I) and eastern Europe, western Europe, and Others (Annex I). Climate change institutions have traditionally added an additional seat for small island developing countries.

12.5.2 Exchanging Benefits for the Right to Use — The transaction at the core of the CDM (Article 12[3]) is described so ambiguously as to leave unanswered a fundamental question: Who finances CDM project activities and what is the relationship between such funding and the extent to which any particular Annex I party can use the resulting CER to offset its commitment? Although Article 12(3)(a) provides that non–Annex I parties are to benefit from project activities and Article 12(3)(b) allows Annex I parties to use the CERs that the project activities generate, there is no direct link between the provision of an investment and the ownership of the offset.

Guidance can be taken from Article 3.12, which provides that CERs can be acquired by one party from another party. This suggests but does not require that a project activity–related investment takes place in exchange for a CER. Indeed, while Article 12.6 leaves open the possibility that the "CDM shall assist in arranging funding of certified project activities *as necessary,*" it is not clear that the CDM will involve the transfer of funds in any traditional sense of public or private project finance.

Explicit references to the need for financial additionality were not included in Article 12. This can be explained in part by the perceptions of some negotiators that private-sector investments, which are expected to generate the bulk of CDM project activities, are by definition additional public sector ODA. Such investments could not, therefore, erode the level of publicly provided development assistance. However, at and since Kyoto at least one delegation, desperate for a means of generating CDM offsets, has proposed to run its climate-related bilateral ODA through the CDM.

Either way the identification of investment tied to particular project activity will clearly help establish the overall additionality of the resulting emissions reduction. Certification of CDM project activities will, after all, depend on proof that reductions in emissions are additional to any that would occur in the of the absence of project activity.

The gap in the transaction between Article 12(3)(a) and 12(3)(b) allows development of a number of proposals that might take the CDM in unanticipated directions. The disjunction between the beneficiary of the investment and the user of the CER raises the possibility that entities may act as intermediaries between investors and hosts to pool funds and build a portfolio of projects involving a variety of hosts. The creation of such financial instruments could introduce liquidity into the system that would allow CERs, or pools thereof, to be held or transferred. Finally, the gap invites discussion as to how CERs might be appropriately shared between an investor and a host.

In what might prove to be the most revolutionary aspect of the CDM, Article 12(9) invites the participation of private and/or public entities (i.e., nonstate actors) into both sides of an Article 12(3) transaction. Proposals by multilateral development banks and both commercial and not-for-profit companies reveal that the nonstate actors are already beginning to position themselves as CDM brokers.[49]

12.5.3 Certification, Auditing, and Verification — Drawing from the experiences and principles established in the AIJ pilot phase, Article 12 recognizes that the key to credible CERs will be the rules, procedures, and principles that will govern the certification of emissions reductions and the auditing and verification of project activities. The principles for emissions reduction certification set out in Article 12(5) will require a return to the fraught political and methodological issues of environmental additionality raised by both the AIJ pilot phase and GEF operations.

The COP/MOP's approach to project activity certification could run the range from laissez-faire to heavily interventionist. Article 12 certainly opens the possibility that a CDM project activity could involve only minimal participation of governments or intergovernment institutions. One scenario, described as ideal by an industry representative in Kyoto, described an Annex I–based parent corporation investing in energy savings in a non–Annex I subsidiary and offsetting the resulting emissions reductions to avoid domestic regulations. Certifying such activities would likely generate a high volume of CDM CERs.

However, this laissez-faire approach runs the risk of undermining the CDM's objective of achieving environmental additionality. It is not clear under these circumstances that the energy efficiency investment would not have otherwise occurred. Furthermore, the absence of any constraints on the emissions of developing countries could lead to substantial leakage of emissions. In a worst-case scenario, the same parent corporation could pay for its energy efficiency investment in one non–Annex I country by switching to cheaper but more polluting processes for a subsidiary in the same or another non–Annex I country. In these circumstances the parent would be able to enjoy an increase in emissions both at home and abroad and suffer no adverse consequences.

At the other end of the spectrum, the CDM certification requirements could seek to be as exacting as the GEF's project cycle. Before a GEF project can claim to have generated a global environmental benefit, a project designer must construct a baseline of domestic activity that would have occurred had GEF

[49] See "Global Carbon Initiative of the World Bank," at http://www.worldbank.org, and plans by the Inter-American Development Bank to establish a pilot CDM program, press release NR-119/98 at http://www.iadb.org.

funding not been provided. To avoid what the GEF describes as the moral hazard that might tempt governments to lower a domestic environmental baseline in order to become eligible for a larger GEF grant, the project baseline must reflect a minimal standard of environmental reasonableness. In other words, the level of emissions reductions credited to a project must be based not just on what would have taken place but also on what should have taken place. Applied to the previous scenario, this would require the parent corporation to demonstrate that its subsidiary was operating in an environmentally reasonable manner before taking credit for emissions reduced through an additional investment.

The GEF's closely regulated approach to project design was demanded primarily by Annex II parties anxious to be reassured that their GEF contributions were being well spent and spent on activities that would not have otherwise occurred. The CDM has the potential to reverse this incentive. If the bulk of financial resources flowing through the CDM are from the private sector, government finance departments will be less concerned with designing rigorous rules. Indeed Annex I countries as a group will have an incentive to lower barriers to project certification, as it will increase the amount of emissions reduction units available to offset their obligations.

Applying high standards for CDM project activity certification by, for example, demanding the same standard of environmental reasonableness from CDM project proponents, as is currently sought from GEF project proponents, has some appeal. Doing so does, however, increase the possibility that the flow of projects might remain limited.

Auditing and verifying a project activity to ensure that it is achieving the emissions reduction units it has promised to its investors is to be carried out by as yet undetermined entities according to modalities and procedures elaborated by the COP/MOP. It seems appropriate that this task be carried out by entities wholly independent of the governments and operational bodies that are designing and implementing the projects. It has been suggested that internationally recognized accounting or consulting firms might perform this function. During the AIJ pilot phase and through similar emissions trading experiments, both private sector and not-for-profit agencies have been developing the expertise and the public profile to put them in a position to play this role.[50]

---

[50]Note the activities of the multinational environmental auditing firm SGS (http://www.sgsgroup.com) and the establishment by the U.S. NGO, the Environmental Defense Fund, of a nonprofit company that will provide comprehensive reporting and tracking of emissions reductions in company-to-company emissions trades. See the Environmental Resources Trust's Web site at http://www.ert.net.

12.5.4 Limitations on the Use of the CDM — The rapid negotiation of Article 12 did not resolve the concerns of all delegations about the equity or the effectiveness of the CDM. This is most clearly indicated by the limitation in Article 12(3)(b) that CERs may only "*contribute to* compliance with a *part of*" Article 3 commitments, as determined by the COP/MOP. Efforts to restrict this part to a specific percentage in the text of the protocol were unsuccessful, and proponents of this flexibility mechanism have indicated that they interpret the provision as being a qualitative guide rather than a quantified cap. Any final decision on the size or character of the limitation will depend on an analysis of the volume of CERs that the CDM is likely to generate, given the transaction costs it might bear, and the extent to which it will have to compete with the protocol's other flexibility mechanisms.

Limitations on the types of project activities

Given the ongoing debate about the CDM, it has been suggested that the COP/MOP develops policies that seek to limit certification, under an elaboration of Article 12(5), to categories of project activities that are agreed in advance to have "real, measurable and long-term benefits." The absence of any mention of sinks in Article 12, in the context of their express inclusion in the parallel language of Article 6, will provide a basis for exploring whether land use change and forestry activities should be excluded from certification until scientific uncertainties associated with those projects are reduced.

Restrictions on participation: Compliance conditionality

As has been discussed, the creation of flexibility mechanisms also creates the possibility of suspending the right to access those mechanisms as a means of encouraging compliance with the protocol's obligations. On the basis of proposals from the United States, such compliance conditionalities were attached to Article 6 (Joint Implementation amongst Annex I Parties).[51]

As the parties begin to review the inconsistencies between the protocol's various flexibility mechanisms, it might prove appropriate to extend similar rules restricting access to the CDM to investors and hosts from parties that are in compliance with all the regime's obligations.

[51] Under Article 6.1(d), an Annex I party is prohibited from acquiring emissions reduction units unless it is in compliance with its inventory and reporting obligations under Articles 5 and 7. Furthermore, should a question arise through the protocol's in-depth review procedures with regard to a party's compliance with Article 6, it may not apply its emissions reduction units until the question is resolved.

Restrictions on timing: Ex post certification

Concerns about the risks associated with some or all of the project activities run through the CDM might be met by allowing CERs only after the project activity has been completed. For example, for an investment in the retooling of a power plant with a 20-year life span, only the actual emissions reduced during the commitment period in question could be offset against that period's assigned amount. There is some basis for this ex post approach in the text of Article 12, which refers to emissions reductions accruing from project activities (suggesting that they must have already occurred to be credited). There will, however, be pressure from investors to offset the full projected value of their investment as soon as possible, perhaps prior to them having fully matured.

12.5.5 Administrative Expenses and Adaptation Costs — A final, revolutionary aspect of the CDM is Article 12(8), which authorizes the COP/MOP to ensure that a share of the proceeds from certified project activities is used both to cover administrative expenses and to assist with adaptation costs. This was the last paragraph of Article 12 to be agreed and was slowed by concerns that it might establish a precedent for the collection by an international body of a tax on private economic activity, usually the exclusive preserve of sovereign states.[52] Similar revenue-raising proposals had been floated in the climate change negotiations before in the context of well-head taxes and bunker fuels taxes and were rejected as radical extensions of supranational authority.

As adopted Article 12(8) leaves open the possibility that expenses and costs can be recovered by national authorities. The article is unclear as to whether "proceeds" is intended to mean financial profits generated by an investment (if any) or some valuation of the CERs generated.

It is furthermore unclear what role the CDM will play in authorizing the expenditure of adaptation funding once it is collected. The parties can anticipate difficult questions as to what kind of projects should be funded in which developing countries. As adaptation funding is always likely to be scarce in the face of an incalculable demand, proposals to "stage" adaptation can be expected.

Both the administrative and the adaptation surcharge raise issues with regard to the CDM's ability to compete with the protocol's other flexibility

---

[52] As will be seen from a comparison of FCCC/CP/1997/CRP.4, Articles 12 and 14 of the Kyoto Protocol, the characterization of how administrative costs could be raised was one of the last parts of the package to be agreed.

mechanisms, which are not, at present, required to cover their costs or to contribute to adaptation. The rate at which CDM proceeds are tapped will need to be set with regard to the elasticity of investors' demand for CERs.

12.6 Conclusion

Since Kyoto the CDM has been the focus of intense interest and speculation among governments and the private sector and among a number of intergovernment and nongovernmental organizations that have seen within the text of Article 12 the potential to further develop or to invent roles for themselves in carrying out its multifaceted functions. Because it holds the aspirations of so many different constituencies, progress in elaborating the details of the CDM might well provide the first indications of the longer term prospects for the protocol as a whole.

Chapter 13

Knowledge and the Environment: Markets with Privately Produced Public Goods

Graciela Chichilnisky

13.1 Privately Produced Public Goods

What do environmental emissions have in common with knowledge? This chapter sees both as *privately produced public goods* [1] and gives conditions for efficient allocation of resources in economies with such goods. These conditions are independent of the units of measurement and extend those of Lindahl, Bowen, and Samuelson for standard public goods. The motivation is to understand efficiency in markets in which new types of items such as knowledge and environmental assets are traded along with standard private goods. Both are public goods in that they are not rival in consumption. However, they are privately produced and thus differ from classical public goods that are produced by governments.

Following Chichilnisky, Heal, and Starrett [1], we consider competitive markets, in which every trader faces the same price for each good. The institutional structure for trading public goods contemplated here is similar to that of the emissions markets for sulfur dioxide (SO_2) that are traded in the Chicago Board of Trade since 1993. The example of global emissions markets is especially interesting. These were created recently by Article 17 of the Kyoto Protocol, where 166 nations explicitly agreed on the creation of such tradable rights among Annex B countries, which are mostly industrial nations (see chapters 11, 12, and 14 of this volume, and the Appendix). These markets were

[1] For the foundations of economies with public goods, the reader is referred to the excellent work of Laffont [4,5] and Varian [6].

formally proposed by the scientists of Columbia's Program on Information and Resources (PIR) to the United Nations Framework Convention on Climate Change in May 1994 and emerged in December 1997 in the Kyoto Protocol (see, e.g., Chichilnisky [2]).

Another example of a privately produced public good is the total amount of knowledge in society. In idealized terms this can be represented by products (e.g., software) that can be duplicated at no cost, so the good is not rival in consumption. Knowledge is often privately produced, thus satisfying the definition of privately produced public good that is provided here. In the case of knowledge, the traders' property rights could be interpreted as rights to use a certain number of licences for knowledge products (e.g., software; see Chichilnisky [3]). In all cases the markets considered here are competitive throughout.

13.2 Markets with Privately Produced Public Goods

We use the model of chapter 3 of this book. A competitive market has $J \geq 2$ traders, one public good denoted $a \in R$, for example, the concentration of carbon dioxide (CO_2) in the atmosphere, and a private good denoted $c \in R$. The property rights of the trader j, $\bar{a}_j$, restrict the amount of carbon that the trader has the right to emit at no cost. More rights can be bought or sold in a competitive market. Each trader produces an amount of the private good $c_j^* \in R$, chosing an input of the public good a_j^* that maximizes profits. Formally,

$$a_j^* = \arg \max_{a_j \in (0,\ \infty)} [\phi_j(a_j) - \pi a_j],$$

where π is the relative price of a, $c_j = \phi_j(a_j)$ and $\partial c_j / \partial a_j < 0$. Private goods are the numeraire, that is $p_c \equiv 1$, and $(\bar{a}_j - a_j)$ represents the amount of carbon emitted by trader j to produce private goods over and above its initial rights. Each trader chooses his or her consumption of private and public goods c_j^*, $a^* \in R^{N+1}$ so as to maximize utility

$$\max_{c_j, a} u_j(c_j,\ a),$$

subject to a budget constraint:

$$c_j + \pi(\bar{a}_j - a_j) = \phi_j(c_j);$$

that is, the value of consumption of private and public goods equals the value of production.

In a competitive market equilibrium all markets clear:

$$\sum_{j=1}^{J} a_j^* = \sum_{j=1}^{J} \bar{a}_j = a^*,$$

and

$$\sum_{j=1}^{J} c_j^* = \sum_{j=1}^{J} \phi_j(a_j^*).$$

13.3 Efficiency Conditions

This section derives efficiency conditions for the allocation and provision of privately produced public goods. The classic Lindahl-Bowen-Samuelson conditions for Pareto efficiency in the supply of classic public goods do not in principle apply here because the public good a is privately produced: Here each producer has different production functions, and maximizes profits.

PROPOSITION 1 Efficiency requires that for every trader j, the sum over all traders of the marginal rates of substitution between the private and the public good should equal the corresponding rate of transformation ϕ_j'. In a competitive market, this rate of transformation must equal the relative price of the privately produced public good.

PROOF. By definition an allocation is Pareto efficient if there is no other feasible allocation that makes everyone as well off and someone strictly better off. By definition, therefore, at such an allocation each trader maximizes his or her utility given the (fixed) levels of utility of all others. Formally, for the J traders, a Pareto-efficient allocation $[c_1, \Sigma_{j=1}^{J} \varphi_j(c_j), \ldots, c_J, \Sigma_{j=1}^{J} \varphi_j(c_j)]$ solves the problem

$$\max_{c_1 \in R^+} \left\{ u_1(c_1, \sum_{j=1}^{J} \varphi_j(c_j)) \right\}$$

$$\text{subject to} \quad \left\{ u_2[c_2, \sum_{j=1}^{J} \varphi_j(c_j)] \right\} = \overline{u_2}, \ldots, \left\{ u_J[c_J, \sum_{j=1}^{J} \varphi_j(c_j)] \right\} = \overline{u_J},$$

where $\varphi(c_j) = \phi_j^{-1}(c_j)$ and $\Sigma_{j=1}^{J} a_j = \Sigma_{j=1}^{J} \phi_j^{-1}(c_j) = a$.

To obtain an optimum, one considers the so-called Lagrangian expression $\{u_1[c_j, \Sigma_{j=1}^{J} \varphi_j(c_j)]\} + \Sigma_{j=2}^{J} \lambda_j \{u_j[c_j, \Sigma_{j=1}^{J} \varphi_j(c_j)]\}$, where the λ_j''s are so-called Lagrangian multipliers, and maximizes the expression

$$\max \sum_{j=1}^{J} \lambda_j \left\{ u_j[c_j, \sum_{j=1}^{J} \varphi_j(c_j)] \right\}, \qquad (13.1)$$

where $\Sigma_{j=1}^{J} a_j = \Sigma_{j=1}^{J} \phi_j^{-1}(c_j) = a$.

Optimizing (13.1), one obtains for each trader j

$$\lambda_j \partial u_j / \partial c_j = -\left(\sum_{j=1}^{J} \lambda_j \partial u_j / \partial a \right) \varphi'_j \tag{13.2}$$

or

$$\lambda_j = (-K) \frac{\varphi'_j}{\partial u_j / \partial c_j}, \tag{13.3}$$

where for all j, K is the same:

$$K = \left(\sum_{j=1}^{J} \lambda_j \partial u_j / \partial a \right).$$

Substituting (13.3) into (13.2), one obtains

$$\lambda_j \partial u_j / \partial c_j = -\left(\sum_{j=1}^{J} -K \varphi'_j \frac{\partial u_j / \partial a}{\partial u_j / \partial c_j} \right) \varphi'_j$$

or

$$\frac{\lambda_j \partial u_j / \partial c_j}{\varphi'_j} = (-K) \left(\sum_{j=1}^{J} \varphi'_j \frac{\partial u_j / \partial a}{\partial u_j / \partial c_j} \right).$$

Substituting φ'_j from (13.2), one obtains

$$\sum_{j=1}^{J} \varphi'_j \frac{\partial u_j / \partial a}{\partial u_j / \partial c_j} = 1 \tag{13.4}$$

or

$$\sum_{j=1}^{J} \frac{\partial u_j / \partial a}{\partial u_j / \partial c_j} = \phi'_j,$$

which is an expression generalizing the Lindahl-Bowen-Samuelson (LBS) conditions for Pareto efficiency in the allocation of *privately* produced public goods. It requires that for each trader j the sum of the marginal rates of substitution should equal the corresponding rate of transformation. Observe that in a competitive market, this equals the relative market price of the privately produced public good but not otherwise, and this condition need not be similar to the LBS condition. This completes the proof. ■

REMARK 1 Observe that expression (13.4) is independent of the units of measurement and it does not depend on the weights λ_j.

References

1. Chichilnisky, G., G. Heal, and D. Starrett. "International Markets with Emissions Trading: Equity and Efficiency." Discussion Paper No. 81, Center for Economic Policy Research, Stanford University, fall 1993. (Chapter 3 of this volume)
2. Chichilnisky, G. "Development and Global Finance: The Case for an International Bank for Environmental Settlements." Discussion Paper No. 10, Office of Development Studies of UNDP, UNESCO, New York, 1997. (Chapter 11 of this volume)
3. Chichilnisky, G. "Markets for Knowledge." Invited lecture at the National Academy of Sciences Forum "Nature and Human Society: The Quest for a Sustainable World," October 27–30, 1997. Published in its proceedings, ed. P. Raven, National Academy of Sciences, Washington, D.C., 2000.
4. Laffont, J.-J. *Effets Externes et Theorie Economique.* Monographies du Seminaire d'Econometrie. Paris: Editions du Centre National de la Recherche Scientifique, 1977.
5. Laffont, J.-J. *Fundamentals of Public Economics.* Cambridge: MIT Press, 1988.
6. Varian, H. *Microeconomic Analysis.* New York: W. W. Norton, 1984.

Chapter 14
A Commentary on the Kyoto Protocol

Raúl Estrada-Oyuela

14.1 Introduction

The Kyoto Protocol to the UN Framework Convention on Climate Change is the product of 30 months of complex negotiations and of a climactic last-minute adoption, so that a good number of its articles and paragraphs need interpretation and further elaboration. In many points agreements were reached on the basis of the "openness" of the drafting and postponement of definitions.

During the next two years a lot of work had to be done on those points through informal gatherings and workshops and intergovernment conferences. Such meetings have already been held by the Royal Institute of International Affairs at Chattam House in London, the Japanese Ministry of Trade and Investment (MITI) in Tokyo, and the OECD and IEA in Paris. This workshop at Columbia University can be considered the fourth in chronological order.

The protocol contains the following:

1. Menu of "mandatory" policies and measures for industrialized countries (Art. 2)
2. Quantified emission limitations and reductions commitments for the same group (Art. 3)
3. A "bubble" provision, especially convenient for the European Union (Art. 4)

"From Kyoto to Buenos Aires: Technology Transfer and Emissions Trading" was chaired by G. Chichilnisky and R. Estrada-Oyuela and hosted by the Italian Academy of Columbia University, April 1998, with the support of the UNESCO chair at Columbia University, UNDP, UNEP, and UNIDO.

4. Requirement of a national system for estimation of emissions (Art. 5)
5. Joint implementation among developed countries with credits (Art. 6)
6. A request for inventories and information on compliance (Art. 7)
7. Mechanism to review information (Art. 8)
8. Provision for the periodical review of the protocol (Art. 9)
9. Indicative policies and measures for all parties (Art. 10)
10. Additional rules for the "financial mechanism" (Art. 11)
11. The "clean developed mechanism" (Art. 12)
12. Rules for the bodies of the protocol, such as the meeting of the parties (Art. 13), the secretariat (Art. 14), the Subsidiary Bodies for Scientific and Technological Advice and for Implementation (Art. 15) and the multilateral consultative process (Art. 16)
13. "Emission trading" (Art. 17; see also Arts. 3.10 and 3.11)
14. An enabling article to adopt noncompliance rules (Art. 18)
15. Rules for the settlement of disputes (Art. 19), amendments (Art. 20), and annexes (Art. 21)
16. Voting rights (Art. 22), designation of the depository (Art. 23), signature, and ratification
17. Accession, acceptance or approval (Art. 24), and entry into force (Art. 25)
18. Reservations (Art. 26), withdrawal (Art. 27), and languages (Art. 28)
19. Two annexes related to Article 3: Annex A, which lists the greenhouse gases controlled by that article and source categories by sectors, and Annex B, which lists the assigned amounts of greenhouse gases for each industrialized country, as percentages of their 1990 emissions.

Analysis and discussion until now have been devoted mainly to quantified commitments and "flexibility mechanisms." However, they do not stand alone and must be understood in the context of the whole protocol.

14.2 Policies and Measures (P&M)

After the definitions in Article 1, the protocol has (Art. 2), a menu of mandatory policies and measures to be implemented by industrialized countries in order to achieve the quantified limitations and reductions of Article 3 and the annexes. The mandatory versus the indicative character of these policies and measures was discussed until Kyoto. The European Union insisted on the mandatory approach, and the United States rejected any kind of references to policies and measures. The final text seems to have little beyond Article 4 of the convention, but it brings additional emphasis on coordination (paragraphs 1.b and 4), and a specific reference to ICAO and IMO (paragraph 2) with the consequent complication of different memberships in those organizations and the protocol.

According to Article 2.3, the parties shall strive to implement policies and measures in such a way as to minimize "adverse effects" on climate change and trade, as well as social, environmental and economic "impacts" on other parties, especially developing country parties and especially on those sensitive to the fossil fuel trade and the small islands states. OPEC countries proposed this paragraph and the similar one in Article 3.14.

14.3 Quantified Emissions Limitation and Reduction (QELROS)

Article 3 establishes the commitment of developed countries to reduce their emissions of greenhouse gases listed in Annex A by at least 5% below the base year in the period 2008–12, in the context of the differentiation of Annex B. Actually, the algebraic addition of reductions and limitations means −5.2%, which might seem a modest target. Its real significance is appreciated when compared with the "business as usual" projected increase of 24% because then it becomes clear that the real reduction will be close to 30%.

14.3.1 Emissions Levels Indicated in Annex B Are Called "Assigned Amounts" — By 2005 each party shall have made "demonstrable progress" in achieving its commitments under the protocol (Art. 3.2), but it has not yet been decided how progress should be demonstrated.

Opting for the "net emissions" approach, Article 3.3 establishes which removals of greenhouse gases can be taken into account. The following paragraph (Art. 3.4) requests information on carbon stocks, provides for the adoption of new modalities to be applied in future "budgets," but also opens the possibility for the immediate application of those modalities. They both enshrine a form of flexibility and important criteria for the implementation of the protocol.

Articles 3.5 and 3.6 create flexibility for the so-called economies in transition in order to select the base year and to implement commitments in other articles.

"Target year" versus "budget" was one of the main issues of negotiation, and finally the "budget" concept was adopted under the name "commitment period." That is defined in Article 3.7, and according to Article 3.9 subsequent periods shall be established in amendments to Annex B. Article 3.8 provides additional flexibility on the base year.

Article 3.10–.13 links flexibility mechanisms and commitments. They provide for additions and subtractions in cases of trading and joint implementation and additions derived from the application of the Clean Development Mechanism (CDM) and from "banking." Trading, joint implementation, and CDM will be the subject of further comments, but "banking" is regulated only by Article 3.13, and the purpose is to allow a party that was able to emit below its "assigned amount" to preserve that portion for the following "commitment period" as it was proposed by the United States.

14.4 Bubbles

Article 4 gives the parties the possibility of creating a "bubble" in order to implement their commitments together. They must notify the secretariat of the protocol of the terms of their agreement. They need not belong to a regional economic integration organization (REIO) to create the bubble, nor need they be neighbors. If they do belong to an REIO, changes in its membership will not alter the commitments. Each party to a bubble shall be responsible for its own level of emissions and if the REIO is a party to the protocol (the European Community is a party to the convention), each country party and the REIO shall be responsible for its levels of emissions.

The bubble concept was also a difficult issue of negotiation and remains a matter to be analyzed. It is well known that individual countries in the European Union have agreed on a "burden sharing" in which some countries are going to reduce emissions, others will keep the 1990 level, and finally a group will increase emissions. However, all of them appear in Annex B with the same percentage of reduction, but after the formal notification of the burden sharing agreement, each country party will be responsible for its commitment under the agreement, not the one in Annex B, and the European Community will also have an established commitment. A discussion on the legal competence of the European Community to achieve its commitment will most probably arise.

14.5 Methodologies

Article 5 requires that each party in Annex I of the convention have a national system for the estimation of emissions and removals before 2008, consistent with methodologies already adopted by the convention bodies. It is necessary to recall that certainty in the estimation of emissions differs significantly between different gases and different sources of the same gas. The "global warming potential" (GWP) concept, utilized by the (IPCC) to compare and add emissions of different gases, is evolving methodologically. The point has special relevance in relation to the additions and subtractions allowed in accordance with flexibility mechanisms.

14.6 Flexibility Mechanisms: Common Issues

(a) Emissions and removals: May emissions be compensated by removals in units of GWP despite the uncertainties on sources and removals?

(b) Gases: May reductions and increases of emissions be exchangeable despite the unlike degree of uncertainty on the GPW of different gases? Or should exchanges be done on the same gas?

(c) Sources: Similar questions as in (b), because the certainty of the estimation of emissions of gases varies with the sources.

(d) Supplemental: Joint implementation (JI), emissions trading (ET), and "certified emission reductions" from CDM shall be "supplemental" to the domestic effort, but a "supplemental" needs to be defined.

(e) Private-sector players are foreseen for JI, trading, and CDM. Should they be regulated in some way?

(f) Will monitoring of the flexibility mechanisms be unified?

14.7 Joint Implementation (JI)

In addition to the common issues of flexibility mechanisms spelled out previously, JI (Art. 6) needs clear guidelines on the baselines that will be used to determine the reduction or removals of greenhouse gases, and that is not easy. In order to participate in JI, it is necessary to have a national system for the estimation of emissions and removals (Art. 5), to report on inventories and on compliance with commitments (Art. 7) and to be free of any question on the implementation raised in the review process (Art. 8).

The possibility of having JI projects based on removals is explicit in Article 6.

14.8 Inventories, Communications, and Reviews

Article 7 provides that parties shall add to their inventories and national communications information related to the implementation of the protocol, and Article 8 institutionalizes the in-depth review process of national communications already in force for parties to the convention. Article 8 adds that the review shall identify potential problems and factors influencing the fulfilment of commitments. The secretariat shall list "questions of implementatrion" to be considered by the meeting of the parties.

Periodical review of the protocol is foreseen by Article 9 and shall be coordinated with the pertinent reviews under the convention.

14.9 All Parties Commitments

Article 10 elaborates on nonquantified commitments for industrialized and developing countries. To identify progress beyond Article 4.1 of the convention requires quite a semantic effort. However, that part of the protocol should be used to promote action from both industrialized and developing countries.

14.10 Financial Mechanism

Article 11 adds the Kyoto Protocol's emphasis to the convention provisions on the financial mechanism.

14.11 Clean Development Mechanism (CDM)

Article 12 defines the Clean Development Mechanism (CDM) that has been called the "Kyoto Surprise." A Clean Development Fund was in previous drafts of the protocol but it had different features. Explanations given in the U.S. Senate show the CDM as JI between developed and developing countries, with participation of the private sector. In addition to common questions related to all flexibility mechanisms, CDM raises specific points:

1. As in the case of JI, it will be necessary to define "baselines."
2. Because recipient countries will not have quantified commitments, it will also be needed to define "reductions in emissions that are additional to any that would occur in the absence of the certified project activity."
3. Another question to solve is, What are "proceeds from certified project activities" (the whole cost of the project, only the donor part of the cost, and so on?). How much will be used to cover "administrative expenses"? How much assistance is needed in adaptation projects?
4. How much are transaction costs increased that way?
5. Article 12 refers to emissions limitation and reduction, not to removals of GHG by sinks, as Article 6 does: May carbon sequestration projects be included in the CDM?

6. Certified emission reductions from 2000 to 2008 can be used in the first commitment period, but the implications of that rule shall be analyzed before the entry into force of the protocol. Most likely the analysis should be preceded by adoption of preliminary rules for the CDM.

14.12 Institutional Economy

Articles 13 to 16 deal with the bodies of the protocol and their procedures, with a tendency to use as far as possible bodies and procedures of the convention.

14.13 Trading

Article 17 mandates the Conference of the parties to define principles, modalities, rules, and guidelines for trading before the entry into force of the protocol. It has to be done for the purposes of verification, reporting, and accounting, and a number of governments have announced that the trading is substantive for their participation in the protocol.

In addition to the general points for all flexibility mechanism described previously in paragraph 14.6, specific aspects of trading will need to be agreed upon:

1. During the debate in Kyoto, the need to establish equity criteria for trading was crystal clear. "Paper tons" and "hot air" were expressions used to indicate the need to verify emission projections from some developed countries willing to participate in trading.
2. The possible role of the private sector in trading needs to be regulated.
3. Some of the drafts used during negotiations not only mentioned private-sector participation but also included references to intermediaries.
4. The possible creation of transaction body for trading is a question opened by a number of proposals, including some from the World Bank and UN Conference on Trade and Development (UNCTAD).

14.14 Noncompliance

Cases of noncompliance shall be addressed by procedures adopted by the meeting of the parties to the protocol, in conformity with Article 16. In fact Articles 7 and 8 of the protocol already have a device to detect and assess noncompliance (see Article 15), and parties in noncompliance will not be able to use the flexibility mechanisms.

14.15 Additions to Annex B

Among other issues in the final clauses, the extreme rigidity established to modify Annexes A and B, complicating the addition of new parties to Annex B even with the consent of the interested party, can be pointed out.

14.16 Volume of Emissions Required to Enter into Force

Last, but not least, the entry into force of the protocol requires 55 parties, including Annex I parties, which accounted for at least 55% of the total carbon dioxide emissions for the base year (Art. 25). That does not give veto to any individual country, but it does give veto to the United States and Russia together. The protocol can enter into force without one of them but not without both of them. Other, similar combinations of parties are possible under the provisions of Article 25.

Appendix

The Kyoto Protocol of the United Nations Framework Convention on Climate Change

The Parties to this Protocol,

Being Parties to the United Nations Framework Convention on Climate Change, hereinafter referred to as "the Convention",

In pursuit of the ultimate objective of the Convention as stated in its Article 2,

Recalling the provisions of the Convention,

Being guided by Article 3 of the Convention,

Pursuant to the Berlin Mandate adopted by decision 1/CP.1 of the Conference of the Parties to the Convention at its first session, Have agreed as follows:

Article 1

For the purposes of this Protocol, the definitions contained in Article 1 of the Convention shall apply. In addition:

1. "Conference of the Parties" means the Conference of the Parties to the Convention.
2. "Convention" means the United Nations Framework Convention on Climate Change, adopted in New York on 9 May 1992.
3. "Intergovernmental Panel on Climate Change" means the Intergovernmental Panel on Climate Change established in 1988 jointly by the World Meteorological Organization and the United Nations Environment Programme.
4. "Montreal Protocol" means the Montreal Protocol on Substances that Deplete the Ozone Layer, adopted in Montreal on 16 September 1987 and as subsequently adjusted and amended.
5. "Parties present and voting" means Parties present and casting an affirmative or negative vote.

6. "Party" means, unless the context otherwise indicates, a Party to this Protocol.
7. "Party included in Annex I" means a Party included in Annex I to the Convention, as may be amended, or a Party which has made a notification under Article 4, paragraph 2(g), of the Convention.

Article 2

1. Each Party included in Annex I, in achieving its quantified emission limitation and reduction commitments under Article 3, in order to promote sustainable development, shall:
   (a) Implement and/or further elaborate policies and measures in accordance with its national circumstances, such as:
      (i) Enhancement of energy efficiency in relevant sectors of the national economy;
      (ii) Protection and enhancement of sinks and reservoirs of greenhouse gases not controlled by the Montreal Protocol, taking into account its commitments under relevant international environmental agreements; promotion of sustainable forest management practices, afforestation and reforestation;
      (iii) Promotion of sustainable forms of agriculture in light of climate change considerations;
      (iv) Research on, and promotion, development and increased use of, new and renewable forms of energy, of carbon dioxide sequestration technologies and of advanced and innovative environmentally sound technologies;
      (v) Progressive reduction or phasing out of market imperfections, fiscal incentives, tax and duty exemptions and subsidies in all greenhouse gas emitting sectors that run counter to the objective of the Convention and application of market instruments;
      (vi) Encouragement of appropriate reforms in relevant sectors aimed at promoting policies and measures which limit or reduce emissions of greenhouse gases not controlled by the Montreal Protocol;
      (vii) Measures to limit and/or reduce emissions of greenhouse gases not controlled by the Montreal Protocol in the transport sector;
      (viii) Limitation and/or reduction of methane emissions through recovery and use in waste management, as well as in the production, transport and distribution of energy;

(b) Cooperate with other such Parties to enhance the individual and combined effectiveness of their policies and measures adopted under this Article, pursuant to Article 4, paragraph 2(e)(i), of the Convention. To this end, these Parties shall take steps to share their experience and exchange information on such policies and measures, including developing ways of improving their comparability, transparency and effectiveness. The Conference of Parties serving as the meeting of the Parties to this Protocol shall, at its first session or as soon as practicable thereafter, consider ways to facilitate such cooperation, taking into account all relevant information.

2. The Parties included in Annex I shall pursue limitation or reduction of emissions of greenhouse gases not controlled by the Montreal Protocol from aviation and marine bunker fuels, working through the International Civil Aviation Organization and the International Maritime Organization, respectively.
3. The Parties included in Annex I shall strive to implement policies and measures under this Article in such a way as to minimize adverse effects, including the adverse effects of climate change, effects on international trade, and social, environmental and economic impacts on other Parties, especially developing country Parties and in particular those identified in Article 4, paragraphs 8 and 9, of the Convention, taking into account Article 3 of the Convention. The Conference of the Parties serving as the meeting of the Parties to this Protocol may take further action, as appropriate, to promote the implementation of the provisions of this paragraph.
4. The Conference of the Parties serving as the meeting of the Parties to this Protocol, if it decides that it would be beneficial to coordinate any of the policies and measures in paragraph 1(a) above, taking into account different national circumstances and potential effects, shall consider ways and means to elaborate the coordination of such policies and measures.

Article 3

1. The Parties included in Annex I shall, individually or jointly, ensure that their aggregate anthropogenic carbon dioxide equivalent emissions of the greenhouse gases listed in Annex A do not exceed their assigned amounts, calculated pursuant to their quantified emission limitation and reduction commitments inscribed in Annex B and in accordance with the provisions of this Article, with a view to reducing

their overall emissions of such gases by at least 5 per cent below 1990 levels in the commitment period 2008 to 2012.

2. Each Party included in Annex I shall, by 2005, have made demonstrable progress in achieving its commitments under this Protocol.
3. The net changes in greenhouse gas emissions by sources and removals by sinks resulting from direct human-induced land-use change and forestry activities, limited to afforestation, reforestation and deforestation since 1990, measured as verifiable changes in carbon stocks in each commitment period, shall be used to meet the commitments under this Article of each Party included in Annex I. The greenhouse gas emissions by sources and removals by sinks associated with those activities shall be reported in a transparent and verifiable manner and reviewed in accordance with Articles 7 and 8.
4. Prior to the first session of the Conference of the Parties serving as the meeting of the Parties to this Protocol, each Party included in Annex I shall provide, for consideration by the Subsidiary Body for Scientific and Technological Advice, data to establish its level of carbon stocks in 1990 and to enable an estimate to be made of its changes in carbon stocks in subsequent years. The Conference of the Parties serving as the meeting of the Parties to this Protocol shall, at its first session or as soon as practicable thereafter, decide upon modalities, rules and guidelines as to how, and which, additional human-induced activities related to changes in greenhouse gas emissions by sources and removals by sinks in the agricultural soils and the land-use change and forestry categories shall be added to, or subtracted from, the assigned amounts for Parties included in Annex I, taking into account uncertainties, transparency in reporting, verifiability, the methodological work of the Intergovernmental Panel on Climate Change, the advice provided by the Subsidiary Body for Scientific and Technological Advice in accordance with Article 5 and the decisions of the Conference of the Parties. Such a decision shall apply in the second and subsequent commitment periods. A Party may choose to apply such a decision on these additional human-induced activities for its first commitment period, provided that these activities have taken place since 1990.
5. The Parties included in Annex I undergoing the process of transition to a market economy whose base year or period was established pursuant to decision 9/CP.2 of the Conference of the Parties at its second session shall use that base year or period for the implementation of their commitments under this Article. Any other Party included in Annex I undergoing the process of transition to a market economy which

has not yet submitted its first national communication under Article 12 of the Convention may also notify the Conference of the Parties serving as the meeting of the Parties to this Protocol that it intends to use an historical base year or period other than 1990 for the implementation of its commitments under this Article. The Conference of the Parties serving as the meeting of the Parties to this Protocol shall decide on the acceptance of such notification.

6. Taking into account Article 4, paragraph 6, of the Convention, in the implementation of their commitments under this Protocol other than those under this Article, a certain degree of flexibility shall be allowed by the Conference of the Parties serving as the meeting of the Parties to this Protocol to the Parties included in Annex I undergoing the process of transition to a market economy.
7. In the first quantified emission limitation and reduction commitment period, from 2008 to 2012, the assigned amount for each Party included in Annex I shall be equal to the percentage inscribed for it in Annex B of its aggregate anthropogenic carbon dioxide equivalent emissions of the greenhouse gases listed in Annex A in 1990, or the base year or period determined in accordance with paragraph 5 above, multiplied by five. Those Parties included in Annex I for whom land-use change and forestry constituted a net source of greenhouse gas emissions in 1990 shall include in their 1990 emissions base year or period the aggregate anthropogenic carbon dioxide equivalent emissions by sources minus removals by sinks in 1990 from land-use change for the purposes of calculating their assigned amount.
8. Any Party included in Annex I may use 1995 as its base year for hydrofluorocarbons, perfluorocarbons and sulphur hexafluoride, for the purposes of the calculation referred to in paragraph 7 above.
9. Commitments for subsequent periods for Parties included in Annex I shall be established in amendments to Annex B to this Protocol, which shall be adopted in accordance with the provisions of Article 21, paragraph 7. The Conference of the Parties serving as the meeting of the Parties to this Protocol shall initiate the consideration of such commitments at least seven years before the end of the first commitment period referred to in paragraph 1 above.
10. Any emission reduction units, or any part of an assigned amount, which a Party acquires from another Party in accordance with the provisions of Article 6 or of Article 17 shall be added to the assigned amount for the acquiring Party.
11. Any emission reduction units, or any part of an assigned amount,

which a Party transfers to another Party in accordance with the provisions of Article 6 or of Article 17 shall be subtracted from the assigned amount for the transferring Party.

12. Any certified emission reductions which a Party acquires from another Party in accordance with the provisions of Article 12 shall be added to the assigned amount for the acquiring Party.
13. If the emissions of a Party included in Annex I in a commitment period are less than its assigned amount under this Article, this difference shall, on request of that Party, be added to the assigned amount for that Party for subsequent commitment periods.
14. Each Party included in Annex I shall strive to implement the commitments mentioned in paragraph 1 above in such a way as to minimize adverse social, environmental and economic impacts on developing country Parties, particularly those identified in Article 4, paragraphs 8 and 9, of the Convention. In line with relevant decisions of the Conference of the Parties on the implementation of those paragraphs, the Conference of the Parties serving as the meeting of the Parties to this Protocol shall, at its first session, consider what actions are necessary to minimize the adverse effects of climate change and/or the impacts of response measures on Parties referred to in those paragraphs. Among the issues to be considered shall be the establishment of funding, insurance and transfer of technology.

Article 4

1. Any Parties included in Annex I that have reached an agreement to fulfil their commitments under Article 3 jointly, shall be deemed to have met those commitments provided that their total combined aggregate anthropogenic carbon dioxide equivalent emissions of the greenhouse gases listed in Annex A do not exceed their assigned amounts calculated pursuant to their quantified emission limitation and reduction commitments inscribed in Annex B and in accordance with the provisions of Article 3. The respective emission level allocated to each of the Parties to the agreement shall be set out in that agreement.
2. The Parties to any such agreement shall notify the secretariat of the terms of the agreement on the date of deposit of their instruments of ratification, acceptance or approval of this Protocol, or accession thereto. The secretariat shall in turn inform the Parties and signatories to the Convention of the terms of the agreement.
3. Any such agreement shall remain in operation for the duration of the commitment period specified in Article 3, paragraph 7.

4. If Parties acting jointly do so in the framework of, and together with, a regional economic integration organization, any alteration in the composition of the organization after adoption of this Protocol shall not affect existing commitments under this Protocol. Any alteration in the composition of the organization shall only apply for the purposes of those commitments under Article 3 that are adopted subsequent to that alteration.
5. In the event of failure by the Parties to such an agreement to achieve their total combined level of emission reductions, each Party to that agreement shall be responsible for its own level of emissions set out in the agreement.
6. If Parties acting jointly do so in the framework of, and together with, a regional economic integration organization which is itself a Party to this Protocol, each member State of that regional economic integration organization individually, and together with the regional economic integration organization acting in accordance with Article 24, shall, in the event of failure to achieve the total combined level of emission reductions, be responsible for its level of emissions as notified in accordance with this Article.

Article 5

1. Each Party included in Annex I shall have in place, no later than one year prior to the start of the first commitment period, a national system for the estimation of anthropogenic emissions by sources and removals by sinks of all greenhouse gases not controlled by the Montreal Protocol. Guidelines for such national systems, which shall incorporate the methodologies specified in paragraph 2 below, shall be decided upon by the Conference of the Parties serving as the meeting of the Parties to this Protocol at its first session.
2. Methodologies for estimating anthropogenic emissions by sources and removals by sinks of all greenhouse gases not controlled by the Montreal Protocol shall be those accepted by the Intergovernmental Panel on Climate Change and agreed upon by the Conference of the Parties at its third session. Where such methodologies are not used, appropriate adjustments shall be applied according to methodologies agreed upon by the Conference of the Parties serving as the meeting of the Parties to this Protocol at its first session. Based on the work of, *inter alia*, the Intergovernmental Panel on Climate Change and advice provided by the Subsidiary Body for Scientific and Technological Advice, the Conference of the Parties serving as the meeting of the Parties to

this Protocol shall regularly review and, as appropriate, revise such methodologies and adjustments, taking fully into account any relevant decisions by the Conference of the Parties. Any revision to methodologies or adjustments shall be used only for the purposes of ascertaining compliance with commitments under Article 3 in respect of any commitment period adopted subsequent to that revision.

3. The global warming potentials used to calculate the carbon dioxide equivalence of anthropogenic emissions by sources and removals by sinks of greenhouse gases listed in Annex A shall be those accepted by the Intergovernmental Panel on Climate Change and agreed upon by the Conference of the Parties at its third session. Based on the work of, *inter alia*, the Intergovernmental Panel on Climate Change and advice provided by the Subsidiary Body for Scientific and Technological Advice, the Conference of the Parties serving as the meeting of the Parties to this Protocol shall regularly review and, as appropriate, revise the global warming potential of each such greenhouse gas, taking fully into account any relevant decisions by the Conference of the Parties. Any revision to a global warming potential shall apply only to commitments under Article 3 in respect of any commitment period adopted subsequent to that revision.

Article 6

1. For the purpose of meeting its commitments under Article 3, any Party included in Annex I may transfer to, or acquire from, any other such Party emission reduction units resulting from projects aimed at reducing anthropogenic emissions by sources or enhancing anthropogenic removals by sinks of greenhouse gases in any sector of the economy, provided that:
   (a) Any such project has the approval of the Parties involved;
   (b) Any such project provides a reduction in emissions by sources, or an enhancement of removals by sinks, that is additional to any that would otherwise occur;
   (c) It does not acquire any emission reduction units if it is not in compliance with its obligations under Articles 5 and 7; and
   (d) The acquisition of emission reduction units shall be supplemental to domestic actions for the purposes of meeting commitments under Article 3.
2. The Conference of the Parties serving as the meeting of the Parties to this Protocol may, at its first session or as soon as practicable thereafter,

further elaborate guidelines for the implementation of this Article, including for verification and reporting.

3. A Party included in Annex I may authorize legal entities to participate, under its responsibility, in actions leading to the generation, transfer or acquisition under this Article of emission reduction units.
4. If a question of implementation by a Party included in Annex I of the requirements referred to in this Article is identified in accordance with the relevant provisions of Article 8, transfers and acquisitions of emission reduction units may continue to be made after the question has been identified, provided that any such units may not be used by a Party to meet its commitments under Article 3 until any issue of compliance is resolved.

Article 7

1. Each Party included in Annex I shall incorporate in its annual inventory of anthropogenic emissions by sources and removals by sinks of greenhouse gases not controlled by the Montreal Protocol, submitted in accordance with the relevant decisions of the Conference of the Parties, the necessary supplementary information for the purposes of ensuring compliance with Article 3, to be determined in accordance with paragraph 4 below.
2. Each Party included in Annex I shall incorporate in its national communication, submitted under Article 12 of the Convention, the supplementary information necessary to demonstrate compliance with its commitments under this Protocol, to be determined in accordance with paragraph 4 below.
3. Each Party included in Annex I shall submit the information required under paragraph 1 above annually, beginning with the first inventory due under the Convention for the first year of the commitment period after this Protocol has entered into force for that Party. Each such Party shall submit the information required under paragraph 2 above as part of the first national communication due under the Convention after this Protocol has entered into force for it and after the adoption of guidelines as provided for in paragraph 4 below. The frequency of subsequent submission of information required under this Article shall be determined by the Conference of the Parties serving as the meeting of the Parties to this Protocol, taking into account any timetable for the submission of national communications decided upon by the Conference of the Parties.

4. The Conference of the Parties serving as the meeting of the Parties to this Protocol shall adopt at its first session, and review periodically thereafter, guidelines for the preparation of the information required under this Article, taking into account guidelines for the preparation of national communications by Parties included in Annex I adopted by the Conference of the Parties. The Conference of the Parties serving as the meeting of the Parties to this Protocol shall also, prior to the first commitment period, decide upon modalities for the accounting of assigned amounts.

Article 8

1. The information submitted under Article 7 by each Party included in Annex I shall be reviewed by expert review teams pursuant to the relevant decisions of the Conference of the Parties and in accordance with guidelines adopted for this purpose by the Conference of the Parties serving as the meeting of the Parties to this Protocol under paragraph 4 below. The information submitted under Article 7, paragraph 1, by each Party included in Annex I shall be reviewed as part of the annual compilation and accounting of emissions inventories and assigned amounts. Additionally, the information submitted under Article 7, paragraph 2, by each Party included in Annex I shall be reviewed as part of the review of communications.
2. Expert review teams shall be coordinated by the secretariat and shall be composed of experts selected from those nominated by Parties to the Convention and, as appropriate, by intergovernmental organizations, in accordance with guidance provided for this purpose by the Conference of the Parties.
3. The review process shall provide a thorough and comprehensive technical assessment of all aspects of the implementation by a Party of this Protocol. The expert review teams shall prepare a report to the Conference of the Parties serving as the meeting of the Parties to this Protocol, assessing the implementation of the commitments of the Party and identifying any potential problems in, and factors influencing, the fulfilment of commitments. Such reports shall be circulated by the secretariat to all Parties to the Convention. The secretariat shall list those questions of implementation indicated in such reports for further consideration by the Conference of the Parties serving as the meeting of the Parties to this Protocol.
4. The Conference of the Parties serving as the meeting of the Parties

to this Protocol shall adopt at its first session, and review periodically thereafter, guidelines for the review of implementation of this Protocol by expert review teams taking into account the relevant decisions of the Conference of the Parties.

5. The Conference of the Parties serving as the meeting of the Parties to this Protocol shall, with the assistance of the Subsidiary Body for Implementation and, as appropriate, the Subsidiary Body for Scientific and Technological Advice, consider:
   (a) The information submitted by Parties under Article 7 and the reports of the expert reviews thereon conducted under this Article; and
   (b) Those questions of implementation listed by the secretariat under paragraph 3 above, as well as any questions raised by Parties.
6. Pursuant to its consideration of the information referred to in paragraph 5 above, the Conference of the Parties serving as the meeting of the Parties to this Protocol shall take decisions on any matter required for the implementation of this Protocol.

Article 9

1. The Conference of the Parties serving as the meeting of the Parties to this Protocol shall periodically review this Protocol in the light of the best available scientific information and assessments on climate change and its impacts, as well as relevant technical, social and economic information. Such reviews shall be coordinated with pertinent reviews under the Convention, in particular those required by Article 4, paragraph 2(d), and Article 7, paragraph 2(a), of the Convention. Based on these reviews, the Conference of the Parties serving as the meeting of the Parties to this Protocol shall take appropriate action.
2. The first review shall take place at the second session of the Conference of the Parties serving as the meeting of the Parties to this Protocol. Further reviews shall take place at regular intervals and in a timely manner.

Article 10

All Parties, taking into account their common but differentiated responsibilities and their specific national and regional development priorities, objectives and circumstances, without introducing any new commitments for Parties not included in Annex I, but reaffirming existing commitments under Article 4, paragraph 1, of the Convention, and continuing to advance the implementation of these commitments in order to achieve sustainable development, taking into account Article 4, paragraphs 3, 5 and 7, of the Convention, shall:

(a) Formulate, where relevant and to the extent possible, cost-effective national and, where appropriate, regional programmes to improve the quality of local emission factors, activity data and/or models which reflect the socio-economic conditions of each Party for the preparation and periodic updating of national inventories of anthropogenic emissions by sources and removals by sinks of all greenhouse gases not controlled by the Montreal Protocol, using comparable methodologies to be agreed upon by the Conference of the Parties, and consistent with the guidelines for the preparation of national communications adopted by the Conference of the Parties;

(b) Formulate, implement, publish and regularly update national and, where appropriate, regional programmes containing measures to mitigate climate change and measures to facilitate adequate adaptation to climate change:

  (i) Such programmes would, *inter alia*, concern the energy, transport and industry sectors as well as agriculture, forestry and waste management. Furthermore, adaptation technologies and methods for improving spatial planning would improve adaptation to climate change; and

  (ii) Parties included in Annex I shall submit information on action under this Protocol, including national programmes, in accordance with Article 7; and other Parties shall seek to include in their national communications, as appropriate, information on programmes which contain measures that the Party believes contribute to addressing climate change and its adverse impacts, including the abatement of increases in greenhouse gas emissions, and enhancement of and removals by sinks, capacity building and adaptation measures;

(c) Cooperate in the promotion of effective modalities for the development, application and diffusion of, and take all practicable steps to promote, facilitate and finance, as appropriate, the transfer of, or access to, environmentally sound technologies, know-how, practices and processes pertinent to climate change, in particular to developing countries, including the formulation of policies and programmes for the effective transfer of environmentally sound technologies that are publicly owned or in the public domain and the creation of an enabling environment for the private sector, to promote and enhance the transfer of, and access to, environmentally sound technologies;

(d) Cooperate in scientific and technical research and promote the maintenance and the development of systematic observation systems and development of data archives to reduce uncertainties related to the climate system, the adverse impacts of climate change and the economic and social consequences of various response strategies, and promote the development and strengthening of endogenous capacities and capabilities to participate in international and intergovernmental efforts, programmes and networks on research and systematic observation, taking into account Article 5 of the Convention;

(e) Cooperate in and promote at the international level, and, where appropriate, using existing bodies, the development and implementation of education and training programmes, including the strengthening of national capacity building, in particular human and institutional capacities and the exchange or secondment of personnel to train experts in this field, in particular for developing countries, and facilitate at the national level public awareness of, and public access to information on, climate change. Suitable modalities should be developed to implement these activities through the relevant bodies of the Convention, taking into account Article 6 of the Convention;

(f) Include in their national communications information on programmes and activities undertaken pursuant to this Article in accordance with relevant decisions of the Conference of the Parties; and

(g) Give full consideration, in implementing the commitments under this Article, to Article 4, paragraph 8, of the Convention.

Article 11

1. In the implementation of Article 10, Parties shall take into account the provisions of Article 4, paragraphs 4, 5, 7, 8 and 9, of the Convention.
2. In the context of the implementation of Article 4, paragraph 1, of the Convention, in accordance with the provisions of Article 4, paragraph 3, and Article 11 of the Convention, and through the entity or entities entrusted with the operation of the financial mechanism of the Convention, the developed country Parties and other developed Parties included in Annex II to the Convention shall:

(a) Provide new and additional financial resources to meet the agreed full costs incurred by developing country Parties in advancing the implementation of existing commitments under Article 4, paragraph 1(a), of the Convention that are covered in Article 10, subparagraph (a); and

(b) Also provide such financial resources, including for the transfer of technology, needed by the developing country Parties to meet the agreed full incremental costs of advancing the implementation of existing commitments under Article 4, paragraph 1, of the Convention that are covered by Article 10 and that are agreed between a developing country Party and the international entity or entities referred to in Article 11 of the Convention, in accordance with that Article.

The implementation of these existing commitments shall take into account the need for adequacy and predictability in the flow of funds and the importance of appropriate burden sharing among developed country Parties. The guidance to the entity or entities entrusted with the operation of the financial mechanism of the Convention in relevant decisions of the Conference of the Parties, including those agreed before the adoption of this Protocol, shall apply *mutatis mutandis* to the provisions of this paragraph.

3. The developed country Parties and other developed Parties in Annex II to the Convention may also provide, and developing country Parties avail themselves of, financial resources for the implementation of Article 10, through bilateral, regional and other multilateral channels.

Article 12

1. A clean development mechanism is hereby defined.
2. The purpose of the clean development mechanism shall be to assist Parties not included in Annex I in achieving sustainable development and in contributing to the ultimate objective of the Convention, and to assist Parties included in Annex I in achieving compliance with their quantified emission limitation and reduction commitments under Article 3.
3. Under the clean development mechanism:

   (a) Parties not included in Annex I will benefit from project activities resulting in certified emission reductions; and

   (b) Parties included in Annex I may use the certified emission reductions accruing from such project activities to contribute to compliance with part of their quantified emission limitation and reduction commitments under Article 3, as determined by the Conference of the Parties serving as the meeting of the Parties to this Protocol.

4. The clean development mechanism shall be subject to the authority and guidance of the Conference of the Parties serving as the meeting of the Parties to this Protocol and be supervised by an executive board of the clean development mechanism.
5. Emission reductions resulting from each project activity shall be certified by operational entities to be designated by the Conference of the Parties serving as the meeting of the Parties to this Protocol, on the basis of:
   (a) Voluntary participation approved by each Party involved;
   (b) Real, measurable, and long-term benefits related to the mitigation of climate change; and
   (c) Reductions in emissions that are additional to any that would occur in the absence of the certified project activity.
6. The clean development mechanism shall assist in arranging funding of certified project activities as necessary.
7. The Conference of the Parties serving as the meeting of the Parties to this Protocol shall, at its first session, elaborate modalities and procedures with the objective of ensuring transparency, efficiency and accountability through independent auditing and verification of project activities.
8. The Conference of the Parties serving as the meeting of the Parties to this Protocol shall ensure that a share of the proceeds from certified project activities is used to cover administrative expenses as well as to assist developing country Parties that are particularly vulnerable to the adverse effects of climate change to meet the costs of adaptation.
9. Participation under the clean development mechanism, including in activities mentioned in paragraph 3(a) above and in the acquisition of certified emission reductions, may involve private and/or public entities, and is to be subject to whatever guidance may be provided by the executive board of the clean development mechanism.
10. Certified emission reductions obtained during the period from the year 2000 up to the beginning of the first commitment period can be used to assist in achieving compliance in the first commitment period.

Article 13

1. The Conference of the Parties, the supreme body of the Convention, shall serve as the meeting of the Parties to this Protocol.

2. Parties to the Convention that are not Parties to this Protocol may participate as observers in the proceedings of any session of the Conference of the Parties serving as the meeting of the Parties to this Protocol. When the Conference of the Parties serves as the meeting of the Parties to this Protocol, decisions under this Protocol shall be taken only by those that are Parties to this Protocol.
3. When the Conference of the Parties serves as the meeting of the Parties to this Protocol, any member of the Bureau of the Conference of the Parties representing a Party to the Convention but, at that time, not a Party to this Protocol, shall be replaced by an additional member to be elected by and from amongst the Parties to this Protocol.
4. The Conference of the Parties serving as the meeting of the Parties to this Protocol shall keep under regular review the implementation of this Protocol and shall make, within its mandate, the decisions necessary to promote its effective implementation. It shall perform the functions assigned to it by this Protocol and shall:
   (a) Assess, on the basis of all information made available to it in accordance with the provisions of this Protocol, the implementation of this Protocol by the Parties, the overall effects of the measures taken pursuant to this Protocol, in particular environmental, economic and social effects as well as their cumulative impacts and the extent to which progress towards the objective of the Convention is being achieved;
   (b) Periodically examine the obligations of the Parties under this Protocol, giving due consideration to any reviews required by Article 4, paragraph 2(d), and Article 7, paragraph 2, of the Convention, in the light of the objective of the Convention, the experience gained in its implementation and the evolution of scientific and technological knowledge, and in this respect consider and adopt regular reports on the implementation of this Protocol;
   (c) Promote and facilitate the exchange of information on measures adopted by the Parties to address climate change and its effects, taking into account the differing circumstances, responsibilities and capabilities of the Parties and their respective commitments under this Protocol;
   (d) Facilitate, at the request of two or more Parties, the coordination of measures adopted by them to address climate change and its effects, taking into account the differing circumstances, responsibilities and capabilities of the Parties and their respective commitments under this Protocol;

(e) Promote and guide, in accordance with the objective of the Convention and the provisions of this Protocol, and taking fully into account the relevant decisions by the Conference of the Parties, the development and periodic refinement of comparable methodologies for the effective implementation of this Protocol, to be agreed on by the Conference of the Parties serving as the meeting of the Parties to this Protocol;

(f) Make recommendations on any matters necessary for the implementation of this Protocol;

(g) Seek to mobilize additional financial resources in accordance with Article 11, paragraph 2;

(h) Establish such subsidiary bodies as are deemed necessary for the implementation of this Protocol;

(i) Seek and utilize, where appropriate, the services and cooperation of, and information provided by, competent international organizations and intergovernmental and non-governmental bodies; and

(j) Exercise such other functions as may be required for the implementation of this Protocol, and consider any assignment resulting from a decision by the Conference of the Parties.

5. The rules of procedure of the Conference of the Parties and financial procedures applied under the Convention shall be applied *mutatis mutandis* under this Protocol, except as may be otherwise decided by consensus by the Conference of the Parties serving as the meeting of the Parties to this Protocol.

6. The first session of the Conference of the Parties serving as the meeting of the Parties to this Protocol shall be convened by the secretariat in conjunction with the first session of the Conference of the Parties that is scheduled after the date of the entry into force of this Protocol. Subsequent ordinary sessions of the Conference of the Parties serving as the meeting of the Parties to this Protocol shall be held every year and in conjunction with ordinary sessions of the Conference of the Parties, unless otherwise decided by the Conference of the Parties serving as the meeting of the Parties to this Protocol.

7. Extraordinary sessions of the Conference of the Parties serving as the meeting of the Parties to this Protocol shall be held at such other times as may be deemed necessary by the Conference of the Parties serving as the meeting of the Parties to this Protocol, or at the written request of any Party, provided that, within six months of the request being communicated to the Parties by the secretariat, it is supported by at least one third of the Parties.

8. The United Nations, its specialized agencies and the International Atomic Energy Agency, as well as any State member thereof or observers thereto not party to the Convention, may be represented at sessions of the Conference of the Parties serving as the meeting of the Parties to this Protocol as observers. Any body or agency, whether national or international, governmental or non-governmental, which is qualified in matters covered by this Protocol and which has informed the secretariat of its wish to be represented at a session of the Conference of the Parties serving as the meeting of the Parties to this Protocol as an observer, may be so admitted unless at least one third of the Parties present object. The admission and participation of observers shall be subject to the rules of procedure, as referred to in paragraph 5 above.

Article 14

1. The secretariat established by Article 8 of the Convention shall serve as the secretariat of this Protocol.
2. Article 8, paragraph 2, of the Convention on the functions of the secretariat, and Article 8, paragraph 3, of the Convention on arrangements made for the functioning of the secretariat, shall apply *mutatis mutandis* to this Protocol. The secretariat shall, in addition, exercise the functions assigned to it under this Protocol.

Article 15

1. The Subsidiary Body for Scientific and Technological Advice and the Subsidiary Body for Implementation established by Articles 9 and 10 of the Convention shall serve as, respectively, the Subsidiary Body for Scientific and Technological Advice and the Subsidiary Body for Implementation of this Protocol. The provisions relating to the functioning of these two bodies under the Convention shall apply *mutatis mutandis* to this Protocol. Sessions of the meetings of the Subsidiary Body for Scientific and Technological Advice and the Subsidiary Body for Implementation of this Protocol shall be held in conjunction with the meetings of, respectively, the Subsidiary Body for Scientific and Technological Advice and the Subsidiary Body for Implementation of the Convention.
2. Parties to the Convention that are not Parties to this Protocol may participate as observers in the proceedings of any session of the subsidiary bodies. When the subsidiary bodies serve as the subsidiary bodies of this Protocol, decisions under this Protocol shall be taken only by those that are Parties to this Protocol.

3. When the subsidiary bodies established by Articles 9 and 10 of the Convention exercise their functions with regard to matters concerning this Protocol, any member of the Bureaux of those subsidiary bodies representing a Party to the Convention but, at that time, not a party to this Protocol, shall be replaced by an additional member to be elected by and from amongst the Parties to this Protocol.

Article 16

The Conference of the Parties serving as the meeting of the Parties to this Protocol shall, as soon as practicable, consider the application to this Protocol of, and modify as appropriate, the multilateral consultative process referred to in Article 13 of the Convention, in the light of any relevant decisions that may be taken by the Conference of the Parties. Any multilateral consultative process that may be applied to this Protocol shall operate without prejudice to the procedures and mechanisms established in accordance with Article 18.

Article 17

The Conference of the Parties shall define the relevant principles, modalities, rules and guidelines, in particular for verification, reporting and accountability for emissions trading. The Parties included in Annex B may participate in emissions trading for the purposes of fulfilling their commitments under Article 3. Any such trading shall be supplemental to domestic actions for the purpose of meeting quantified emission limitation and reduction commitments under that Article.

Article 18

The Conference of the Parties serving as the meeting of the Parties to this Protocol shall, at its first session, approve appropriate and effective procedures and mechanisms to determine and to address cases of non-compliance with the provisions of this Protocol, including through the development of an indicative list of consequences, taking into account the cause, type, degree and frequency of non-compliance. Any procedures and mechanisms under this Article entailing binding consequences shall be adopted by means of an amendment to this Protocol.

Article 19

The provisions of Article 14 of the Convention on settlement of disputes shall apply *mutatis mutandis* to this Protocol.

Article 20

1. Any Party may propose amendments to this Protocol.
2. Amendments to this Protocol shall be adopted at an ordinary session of the Conference of the Parties serving as the meeting of the Parties to this Protocol. The text of any proposed amendment to this Protocol shall be communicated to the Parties by the secretariat at least six months before the meeting at which it is proposed for adoption. The secretariat shall also communicate the text of any proposed amendments to the Parties and signatories to the Convention and, for information, to the Depositary.
3. The Parties shall make every effort to reach agreement on any proposed amendment to this Protocol by consensus. If all efforts at consensus have been exhausted, and no agreement reached, the amendment shall as a last resort be adopted by a three-fourths majority vote of the Parties present and voting at the meeting. The adopted amendment shall be communicated by the secretariat to the Depositary, who shall circulate it to all Parties for their acceptance.
4. Instruments of acceptance in respect of an amendment shall be deposited with the Depositary. An amendment adopted in accordance with paragraph 3 above shall enter into force for those Parties having accepted it on the ninetieth day after the date of receipt by the Depositary of an instrument of acceptance by at least three fourths of the Parties to this Protocol.
5. The amendment shall enter into force for any other Party on the ninetieth day after the date on which that Party deposits with the Depositary its instrument of acceptance of the said amendment.

Article 21

1. Annexes to this Protocol shall form an integral part thereof and, unless otherwise expressly provided, a reference to this Protocol constitutes at the same time a reference to any annexes thereto. Any annexes adopted after the entry into force of this Protocol shall be restricted to lists, forms and any other material of a descriptive nature that is of a scientific, technical, procedural or administrative character.
2. Any Party may make proposals for an annex to this Protocol and may propose amendments to annexes to this Protocol.
3. Annexes to this Protocol and amendments to annexes to this Protocol shall be adopted at an ordinary session of the Conference of the Parties serving as the meeting of the Parties to this Protocol. The text of any

proposed annex or amendment to an annex shall be communicated to the Parties by the secretariat at least six months before the meeting at which it is proposed for adoption. The secretariat shall also communicate the text of any proposed annex or amendment to an annex to the Parties and signatories to the Convention and, for information, to the Depositary.

4. The Parties shall make every effort to reach agreement on any proposed annex or amendment to an annex by consensus. If all efforts at consensus have been exhausted, and no agreement reached, the annex or amendment to an annex shall as a last resort be adopted by a three-fourths majority vote of the Parties present and voting at the meeting. The adopted annex or amendment to an annex shall be communicated by the secretariat to the Depositary, who shall circulate it to all Parties for their acceptance.
5. An annex, or amendment to an annex other than Annex A or B, that has been adopted in accordance with paragraphs 3 and 4 above shall enter into force for all Parties to this Protocol six months after the date of the communication by the Depositary to such Parties of the adoption of the annex or adoption of the amendment to the annex, except for those Parties that have notified the Depositary, in writing, within that period of their non-acceptance of the annex or amendment to the annex. The annex or amendment to an annex shall enter into force for Parties which withdraw their notification of non-acceptance on the ninetieth day after the date on which withdrawal of such notification has been received by the Depositary.
6. If the adoption of an annex or an amendment to an annex involves an amendment to this Protocol, that annex or amendment to an annex shall not enter into force until such time as the amendment to this Protocol enters into force.
7. Amendments to Annexes A and B to this Protocol shall be adopted and enter into force in accordance with the procedure set out in Article 20, provided that any amendment to Annex B shall be adopted only with the written consent of the Party concerned.

Article 22

1. Each Party shall have one vote, except as provided for in paragraph 2 below.
2. Regional economic integration organizations, in matters within their competence, shall exercise their right to vote with a number of votes

equal to the number of their member States that are Parties to this Protocol. Such an organization shall not exercise its right to vote if any of its member States exercises its right, and vice versa.

Article 23

The Secretary-General of the United Nations shall be the Depositary of this Protocol.

Article 24

1. This Protocol shall be open for signature and subject to ratification, acceptance or approval by States and regional economic integration organizations which are Parties to the Convention. It shall be open for signature at United Nations Headquarters in New York from 16 March 1998 to 15 March 1999. This Protocol shall be open for accession from the day after the date on which it is closed for signature. Instruments of ratification, acceptance, approval or accession shall be deposited with the Depositary.
2. Any regional economic integration organization which becomes a Party to this Protocol without any of its member States being a Party shall be bound by all the obligations under this Protocol. In the case of such organizations, one or more of whose member States is a Party to this Protocol, the organization and its member States shall decide on their respective responsibilities for the performance of their obligations under this Protocol. In such cases, the organization and the member States shall not be entitled to exercise rights under this Protocol concurrently.
3. In their instruments of ratification, acceptance, approval or accession, regional economic integration organizations shall declare the extent of their competence with respect to the matters governed by this Protocol. These organizations shall also inform the Depositary, who shall in turn inform the Parties, of any substantial modification in the extent of their competence.

Article 25

1. This Protocol shall enter into force on the ninetieth day after the date on which not less than 55 Parties to the Convention, incorporating Parties included in Annex I which accounted in total for at least 55 per cent of the total carbon dioxide emissions for 1990 of the Parties included in Annex I, have deposited their instruments of ratification, acceptance, approval or accession.
2. For the purposes of this Article, "the total carbon dioxide emissions for 1990 of the Parties included in Annex I" means the amount communi-

cated on or before the date of adoption of this Protocol by the Parties included in Annex I in their first national communications submitted in accordance with Article 12 of the Convention.

3. For each State or regional economic integration organization that ratifies, accepts or approves this Protocol or accedes thereto after the conditions set out in paragraph 1 above for entry into force have been fulfilled, this Protocol shall enter into force on the ninetieth day following the date of deposit of its instrument of ratification, acceptance, approval or accession.
4. For the purposes of this Article, any instrument deposited by a regional economic integration organization shall not be counted as additional to those deposited by States members of the organization.

Article 26

No reservations may be made to this Protocol.

Article 27

1. At any time after three years from the date on which this Protocol has entered into force for a Party, that Party may withdraw from this Protocol by giving written notification to the Depositary.
2. Any such withdrawal shall take effect upon expiry of one year from the date of receipt by the Depositary of the notification of withdrawal, or on such later date as may be specified in the notification of withdrawal.
3. Any Party that withdraws from the Convention shall be considered as also having withdrawn from this Protocol.

Article 28

The original of this Protocol, of which the Arabic, Chinese, English, French, Russian and Spanish texts are equally authentic, shall be deposited with the Secretary-General of the United Nations.

DONE at Kyoto this eleventh day of December one thousand nine hundred and ninety-seven.

IN WITNESS WHEREOF the undersigned, being duly authorized to that effect, have affixed their signatures to this Protocol on the dates indicated.

Annex A

Greenhouse gases

Carbon dioxide (CO_2)
Methane (CH_4)

Nitrous oxide (N_2O)
Hydrofluorocarbons (HFCs)
Perfluorocarbons (PFCs)
Sulphur hexafluoride (SF_6)

Sectors/source categories

Energy
- Fuel combustion
  - Energy industries
  - Manufacturing industries and construction
  - Transport
  - Other sectors
  - Other
- Fugitive emissions from fuels
  - Solid fuels
  - Oil and natural gas
  - Other

Industrial processes
- Mineral products
- Chemical industry
- Metal production
- Other production
- Production of halocarbons and sulphur hexafluoride
- Consumption of halocarbons and sulphur hexafluoride
- Other

Solvent and other product use

Agriculture
- Enteric fermentation
- Manure management
- Rice cultivation
- Agricultural soils
- Prescribed burning of savannas
- Field burning of agricultural residues
- Other

Waste
- Solid waste disposal on land
- Wastewater handling
- Waste incineration
- Other

Annex B

Party	Quantified emission limitation or reduction commitment (percentage of base year or period)
Australia	108
Austria	92
Belgium	92
Bulgaria*	92
Canada	94
Croatia*	95
Czech Republic*	92
Denmark	92
Estonia*	92
European Community	92
Finland	92
France	92
Germany	92
Greece	92
Hungary*	94
Iceland	110
Ireland	92
Italy	92
Japan	94
Latvia*	92
Liechtenstein	92
Lithuania*	92
Luxembourg	92
Monaco	92
Netherlands	92
New Zealand	100
Norway	101
Poland*	94
Portugal	92
Romania*	92
Russian Federation*	100
Slovakia*	92
Slovenia*	92
Spain	92
Sweden	92
Switzerland	92
Ukraine*	100
United Kingdom of Great Britain and Northern Ireland	92
United States of America	93

* Countries that are undergoing the process of transition to a market economy.

Contributors

Graciela Chichilnisky holds the UNESCO Chair in Mathematics and Economics and is professor of statistics at Columbia University. She taught previously at Harvard, Essex, and Stanford Universities. The director of Columbia's Program on Information and Resources and, since 1998, its Center for Risk Management, Chichilnisky was an early proponent of global carbon emissions markets and played a critical role in their introduction in the Kyoto Protocol. In 1995 she proposed to the World Bank the creation of an International Bank for Environmental Settlements to foster equity and efficiency in global emissions markets. She is the author of the concept of "basic needs" and has published about 180 scientific articles and eleven books, including *Oil in the International Economy* (Oxford University Press) and *Topology and Markets* (American Mathematical Society). She views environmental markets and knowledge markets as key economic developments in the coming century.

Contact Information
Professor Graciela Chichilnisky
Program on Information and Resources
Columbia Center for Risk Management
Columbia University
405 Low Library
535 W. 116th Street
New York, NY 10027
Phone: (212) 854-7275
E-mail: gc9@columbia.edu

Raúl Estrada-Oyuela, born in Buenos Aires, Argentina, is a career diplomat. He graduated in law from the National University of Buenos Aires, was posted in Washington, D.C., Vienna, Brasilia, and Santiago and was the Argentinian ambassador to the People's Republic of China from 1994 to 1997. He attended several sessions of the United Nations General Assembly and participated in numerous international conferences, particularly on environmental issues. He was chairman of the Intergovernmental Negotiating Committee for a Framework Convention on Climate Change (INC/FCCC), created by United Nations General Assembly Resolution 45/212, chairman of the Ad Hoc Group on the Berlin Mandate (AGBM), and chairman of the Committee of the Whole of the Kyoto Conference, where the Protocol was adopted. In 1998 was appointed Distinguished Lecturer on the Global Community and Its Challenges at the Institute for International Studies, Stanford University, and visiting professor, Program on Information and Resources, Columbia University, 1999–2000.

Contact Information

Ambassador Raúl Estrada-Oyuela
La Pampa 1121, 6°
Buenos Aires 1428
Argentina
Phone: (5411) 4786.7255
Fax: (5411) 4322.1593
E-mail: eoy@ciudad.com.ar

Laurent Gilotte graduated from the Ecole Centrale de Paris in 1992. He has been a research assistant at the CIRED and at IIASA (International Institute for Applied System Analysis, Vienna); he has also worked at the World Health Organization (Regional Office for Europe, Copenhagen) as coordinator of the economics topic for the 3rd ministerial conference on Environment and Health. His main field of interest in environmental economics is climate change.

Contact Information

Laurent Gilotte
19 rue Paul Langevin
F-92160 Antony
France
E-mail: laurent.gilotte@voila.fr

Jean-Charles Hourcade is director of research at the Centre National de la Recherche Scientifique (CNRS) and acting director of the CIRED (Centre International de Recherches sur l'Environnement et le Développement), a laboratory belonging to École des Hautes Études en Sciences Sociales, Paris

(E.H.E.S.S.). Since 1990, he has contributed to coordinating French research and advising governmental agencies in charge of climate change. He was the convening lead author of chapters for the Second (1996) Assessment Report (SAR) of the IPCC and is continuing with the Third Assessment Report (TAR). He takes part in various European networks within the EU and the Energy Modeling Forum (EMF), Stanford University.

Contact Information
Jean-Charles Hourcade, Director
CIRED CNRS/EHESS
Jardin Tropical 45bis
Av. de la Belle Gabrielle
94736 NOGENT/MARNE CEDEX
Phone: (33-1) 43.94.73.73
Fax: (33-1) 43.94.73.70
E-mail: hourcade@centre-cired.fr

Joaquim Oliveira Martins is senior economist at the Economics Department of the Organisation for Economic Co-operation and Development (OECD). Previously, he was research fellow at the Centre d'Études Prospectives et d'Informations Internationales (CEPII) Paris. He holds graduate degrees in econometrics and mathematical economics from the University of Paris-I, Panthéon-Sorbonne. His main areas of research have been international trade under imperfect competition, global environmental problems and, more recently, transition economics. He has published several articles, books, and book articles on these topics. He is coauthor of the OECD's GREEN model, a world computable general equilibrium model to assess the costs of reducing CO_2 emissions.

Contact Information
Joaquim Oliveira Martins
OECD-Economics Department
2, Rue André Pascal
75775 PARIS CEDEX 16, FRANCE
Phone: (33-1) 45.24.88.53
Fax: (33-1) 45.24.91.75
E-mail: Joaquim.Oliveira@oecd.org

Andrea Prat is assistant professor of economics at Tilburg University in the Netherlands. In 1997 he received his Ph.D. from Stanford University, where he wrote a thesis on organization theory. His current research focuses on applications of game theory and organization theory to political institutions.

Contact Information
Professor Andrea Prat
B923
Department of Econometrics and Center for Economic Research
PO Box 90153
Tilburg University
5000 LE Tilburg
The Netherlands
Phone: (31-13) 466.27.53 or 466.24.30
Fax: (31-13) 466.32.80
E-mail: a.prat@kub.nl
http://www.center.kub.nl/staff/prat

David Starrett is professor emeritus of economics at Stanford University, at which school he received his Ph.D. in 1968. He also has taught at Harvard and has held visiting appointments at a number of schools, including the London School of Economics, Oxford University, and the Australian National University. He has written on a variety of subjects, mostly concentrated in the areas of economic theory and public economics. He is author of a book on the latter subject, *Foundations of Public Economics* (Cambridge University Press, 1988). Recently his interests have turned to environmental economics and in particular its interface with ecology. Currently he resides in Ely, Nevada.

Contact Information
Professor David Starrett
PO Box 150821
East Ely, NV 89315
Phone and fax: (775) 289-7961
E-mail: davstarret@aol.com

Peter H. Sturm received a Ph.D. from Yale University in 1974. He is currently visiting scholar, International Monetary Fund, Research Department (since 1998). He has held previous positions as: Head of Division (Country Studies, Growth Studies, Macroeconomic and Systems Analysis, Money and Finance Divisions), Organization for Economic Cooperation and Development, Economics Department (1995–98, 1989–94, 1980–87, 1974–79); visiting associate professor, Massey University (New Zealand, 1994–95); special adviser, New Zealand Treasury (1987–89); visiting professor, University of British Columbia (1979–80); assistant professor, Washington University (1972–74); and research fellow, The Brookings Institution (1971–72). His current interests are labor economics and public sector economics.

Contact Information
Peter Sturm
c/o International Monetary Fund
Research Department
700 19th Street
Washington, DC 20431
USA

Jacob Werksman is a senior lawyer at the Foundation for International Environmental Law and Development (FIELD), which provides legal assistance to governments and intergovernmental and nongovernmental organizations on all aspect of international law and sustainable development. Since 1990 he has provided legal assistance to the Alliance of Small Island States (AOSIS) in support of the ongoing development and implementation of the 1992 United Nations Framework Convention on Climate Change and its Kyoto Protocol. At SOAS, University of London, Mr. Werksman lectures in international economic law. Before joining FIELD in 1992, he practiced energy and environmental regulatory law in California, where he remains an active member of the state bar. He holds degrees from Columbia University (A.B. 1986, English Literature); the University of Michigan (Juris Doctor, cum laude, 1990); and the University of London (LLM, 1993, Public International Law).

Contact Information
Jacob Werksman
46–47 Russell Square
London WC1B 4JP
U.K.
E-mail: jw18@soas.ac.uk

Yun Lin has done work in areas such as markets with tradable emission permits, growth of economies with knowledge, and exhaustible natural resource inputs. He graduated from Columbia University in 1996, and spent the next year and half at the Program on Information and Resources at Columbia University as a UNESCO Research Fellow. He joined The Chase Manhattan Bank in 1998 to work on credit card risk management.

Contact Information
Dr. Yun Lin
The Chase Manhattan Bank
100 Duffy Avenue, 4H/1
Hicksville, NY 11801
Phone: (516) 934-2361
Fax: (516) 934-6513
E-mail: Yun.Lin@Chase.com

Index